ONE DAY AT A TIME

The Story of Ainsworth Lumber

ONE DAY AT A TIME

The Story of Ainsworth Lumber

PAUL LUFT

Published in Canada by:
Ainsworth Lumber Co. Ltd.
Bentall Four
3194 – 1055 Dunsmuir Street
Vancouver, British Columbia
Canada V7X 1L3

No part of this book may be reproduced in any form, or by any electronic, mechanical or other means, without written permission from the publisher.

Copyright © 2007, Ainsworth Lumber Co. Ltd.

First edition, 2007
All rights reserved

Printed and bound in Canada

Cover and chapter title pages by Signals and Paul Luft; page design and composition by Paul Luft; editing by Marial Shea and Barbara Tomlin; index by Audrey McClellan.

Library and Archives Canada Cataloging in Publication

Luft, Paul
    One day at a time : the story of Ainsworth Lumber / Paul Luft.

ISBN 0-9781322-0-3

1. Ainsworth Lumber Co.—History. I. Ainsworth Lumber Co. II. Title.

HD9764.C34A35 2006   338.7'634980971   C2006-904008-7

Photograph Credits: *100 Mile House Free Press*, pp. 60, 233, 235; Al Owen, p. 227; Al Smith, pp. 97, 192, 196, 197, 200, 206, 208, 220, 260; Bob McCormack, pp. 140, 199; Demac Mgt. Ltd., p. 247; Dick Sellars, p. 194; Doug White, pp. 170, 184; Fibreco, p. 122; Jack Lefferson, pp. 59, 67, 71, 79, 150, 154, 201, 270; Omer Gosselin, pp. 138, 141, 146, 181, 194, 231; Robin Nadin, pp. 96, 131, 134, 135, 142, 183, 184, 215, 242; Ross Marks, pp. 60, 88-89, 130; Steelworkers Local 1-80 Archives Special Collections, pp. 28, 31, 33, 35, 43; Tom Schaff, pp. 166, 167, 178, 179, 180; all others, Ainsworth family and Ainsworth Lumber.

Reprints of *Vancouver Sun*, *100 Mile House Free Press* and *Grande Prairie Herald-Tribune* used with permission. Ray Williston passage from Ken Drushka's *Tie Hackers to Timber Harvesters* (Harbour Publishing) used with permission. B.C. map template courtesy Roger Handling, Terra Firma Digital Arts.

CONTENTS

FOREWORD — 7

INTRODUCTION — 9

Chapter 1
PRAIRIE ROOTS — 10

Chapter 2
CORK BOOTS AND CHAINSAWS — 24

Chapter 3
NORTH TO THE CARIBOO — 44

Chapter 4
LODGEPOLE PINE PIONEERS — 80

Chapter 5
AGAINST THE GRAIN — 124

Chapter 6
READY FOR TAKE-OFF — 158

Chapter 7
MAKING THE GRADE — 186

Chapter 8
BETTING ON OSB — 226

Chapter 9
A NEW ERA — 264

FOREWORD

We started Ainsworth Lumber more than 50 years ago. Looking back, it's hard to believe how much has occurred over the decades. First, there was growing up on the Canadian prairies during the Great Depression, then logging on Vancouver Island and in the Fraser Valley area, then moving our family north to the Cariboo in 1952 with a portable sawmill. That's when our business really began.

We felt it was important to share this story as a way of preserving Ainsworth's history and acknowledging some of the people who have helped make our company the success it is. Because family has always been very important to us — and has been a big part of our success — this book is also a family history. Having been involved in the forest industry for so long, we have many memories of people and events along the way. Although not all our memories are what you'd call pleasant, it wouldn't be exaggerating to say that facing and overcoming our share of obstacles has got us to where we are today.

Our business philosophy in the early days was just to survive. We made moves simply because we had to. At times, we were driven by the competition to do things differently than we wanted, but in the end, we succeeded.

With our long and lively history, there are obviously far more stories and events than could reasonably be included in one book. Nevertheless, we think the following pages provide a thorough account of Ainsworth Lumber's development and growth, as told by quite a few voices. We've been fortunate to have many dedicated and talented employees over the years, and some of their stories are included here.

Our company's reputation and success are very gratifying to us, and we are especially happy that we've been able to keep our family involved — something we're told is a rare thing.

Looking back, and looking ahead to an exciting future, everyone at Ainsworth can be a little bit proud of what we've all accomplished.

David and Susan Ainsworth
Founders, Ainsworth Lumber Co. Ltd.

ONE DAY AT A TIME

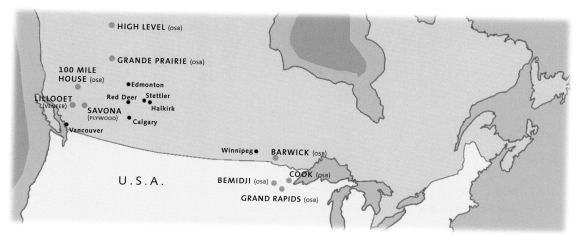

INTRODUCTION

When David Ainsworth set up his portable sawmill in the backwoods of British Columbia, the future extended only as far as his next contract. The Interior's rough-and-tumble forest industry presented formidable challenges. Merely surviving was a constant struggle.

Ainsworth Lumber did far more than survive, however.

What follows is the remarkable story of how a bush mill operation with a handful of employees became one of North America's leading companies in the forest products industry. Ainsworth Lumber is a family business built from the ground up in true pioneer fashion. Every step forward was hard-won, with no shortcuts or deep pockets. In the face of tough conditions and even tougher competitors, founders David and Susan Ainsworth relied on their prodigious work ethic, perseverance and honest business dealings to get them through the day.

Their success was by no means guaranteed. During the company's formative years especially, numerous equipment breakdowns and a host of other minor disasters could easily have put them out of business — and almost did.

As the company grew, so did the risks and challenges.

Innovation plays a central role in the story of Ainsworth Lumber, stretching all the way back to the company's inception.

When their early supplies of Douglas fir grew scarce, David turned his attention to producing premium-grade studs using lodgepole pine — despite the fact that pine was considered a "weed species" at the time, and the market for pine studs was virtually non-existent. The company's history is full of such advances, each one largely a response to whatever obstacle happened to be blocking its path.

David's passion for flying, meanwhile, provides an important backdrop to the company's development, with nearly four decades of accomplished piloting.

Ainsworth's story is also infused with the spirit of entrepreneurship. Starting with David and carried forward by two successive generations, the ability to create opportunities and act decisively has long characterized this family-controlled enterprise.

By the mid-1990s, a shift to the production of oriented strand board marked an enormous change for the company. Ainsworth Lumber has since grown to include nine divisions located in several provinces and the United States, with a talented workforce numbering over 1,800 men and women. It is now the fourth-largest producer of OSB in the world.

But the story doesn't end there.

Despite over half a century in business, Ainsworth finds itself in the midst of an exciting new era, with a wealth of opportunity on the horizon.

As the company moves forward, the inevitable challenges and successes will undoubtedly be taken in stride, just as they were by Ainsworth's founders, one day at a time.

Chapter 1

PRAIRIE ROOTS

Founders of Ainsworth Lumber, David and Susan Ainsworth.

David and Susan Ainsworth had come full circle. Now in their mid-eighties and living in Vancouver, the couple had travelled back to the Cariboo on Canada Day in 2006 to unveil Ainsworth's new forestry display at the 108 Mile Heritage Site. Along with the chainsaws and logging truck featured in the roofed enclosure was the portable sawmill that sparked the creation of the Ainsworth Lumber Company back in the early 1950s.

That sawmill's importance wasn't lost on all the well-wishers who turned out for the ribbon cutting, many of whom were former employees long since retired. The sawmill display symbolizes the humble beginnings of a B.C. forestry company that has grown to prominence in markets around the world.

For those in attendance, David and Susan's presence on that sunny day in July provided a tangible link to the company's origins — and the generations that followed. But the full Ainsworth story stretches back further in time and across the Atlantic, where David's father first heard the call of "The Wondrous West."

George and Ellen Ainsworth were married in 1904 in Fulbeck, Lincolnshire.

At the end of the 19th century, pamphlets by the millions were printed in a dozen languages and sent abroad to entice homesteaders to settle Canada's western provinces — a "Land of Opportunity" where "Prosperity Follows Settlement." Unknown to the prospective settlers, the slick advertisements created by the federal government and the Canadian Pacific Railway often exaggerated the region's attractions or neglected to mention the hardships that newcomers would encounter. But the advertising proved effective. Between 1895 and 1914, the western settlement campaign attracted over one million new residents, many of them to rural Alberta.

Back in London, England, a 33-year-old farrier named George Edwin Ainsworth was one of the multitudes who heard the call of the west. For him, the prospect of a 160-acre homestead clinched the deal.

With Canada in his sights, George married 21-year-old Ellen Shelbourne on February 20, 1904, and 10 months later they were on a steamship bound for Halifax. For Ellen, the crossing was a nightmare. She was seasick the entire time.

George Ainsworth enjoyed the trip immensely; pulling up stakes for Canada was the fulfillment of his dreams.

In those days, newly arrived immigrants to Canada travelled west on "colonists' trains." During the long, 10-day journey, the railcars offered little comfort beyond hard benches that folded down into bunks.

The couple made their way to Red Deer, Alberta, and began their search for a suitable homestead site almost immediately. Under the deal offered by the Canadian government, immigrants to Canada's northwest were given a 160-acre parcel of land for $10. The homesteader had to live on the property for six months of the year, build a dwelling worth at least $300, and break 10 acres of sod each year for three consecutive years. They called it "proving up" the land. If all the requirements were met, full title was granted. Some 60 million acres were made available under this Dominion Land Policy.

Setting out in a horse and buggy, George and Ellen followed a trail east of Red Deer to a point three miles north of what was known as the Higgins stopping house, which was just north of Gadsby.

For Ellen Ainsworth, still reeling from her trans-Atlantic crossing, it must have been a day she'd rather forget. En route to their prospective homesteading site, they had to cross the Red Deer River in the buggy. Although visitors to the area would now find the Content Bridge at this particular ford in the river, there was no such convenience there at the time. As the horse pulled them out into the river's rushing current, the high water quickly came right up into the wagon. A distraught Ellen, holding her skirts up on the dash, was certain they'd be swept away. Fortunately, the horse was familiar with the crossing and pulled them through without incident.

That evening the couple set up camp near the banks of the river. No sooner had they started a fire for their meal when sparks ignited the dry mat of grass surrounding them, sometimes referred to as "prairie wool." With the wind blowing and the flames now

George Ainsworth at his blacksmith shop in rural Alberta. Ellen Ainsworth is seated in the buggy.

consuming ever-larger swaths of dead grass, George and Ellen grabbed whatever they could to put out the fire, starting with a long-handled frying pan. They managed to beat down the flames with a couple of wet sacks, but not before they nearly succumbed to exhaustion.

The newcomers eventually found a site and registered their 160-acre homestead, but they spent the following winter in Red Deer, where George plied his trade as a farrier. The following summer a neighbour ploughed a fireguard and enough sod for a house — the sod was known as "Canadian brick" — although their initial few months on the homestead were spent in a tent.

George, meanwhile, opened a blacksmith shop at the Higgins stopping house. He worked there three days a week during the summer months, and in the winter he and Ellen would move to Calgary, where he worked for the CPR in the Ogden yards. Ellen was employed in the boarding house where they stayed, cleaning and cooking. These winter sojourns to the city weren't unusual at the time; in fact, they were pretty much a necessity. The tents and rustic houses of the early homesteaders offered little comfort when temperatures plummeted.

Ellen and George Ainsworth.

Despite the hardships, George and Ellen did manage to spend one entire year on that homestead. Unfortunately, it was the severest winter the region had ever seen. Cattle starved and froze to death, and a number of area residents lost their lives in blizzards. The snow was so deep that George had to fashion crude cross-country skis out of barrel staves just to get around.

The former Londoner was discovering firsthand just how optimistic those homesteading advertisements really were.

It was early in 1909 that the village of Halkirk, named after a town in Scotland, was officially established by the CPR. The railway still owned huge tracts of land throughout the west. When the CPR began selling property in the village, George felt his prospects were better as a full-time blacksmith than as an inexperienced farmer. He decided to sell the homestead and bought two lots in town: one for a house and one for his own blacksmith shop.

George's boyhood apprenticeship back in England, as well as his subsequent work experience, had prepared him well for the rigours of operating his own smithy. George's father had been a boilermaker, so working with steel was already in the Ainsworth blood.

As a lad of 13, George had joined a small group of boys for blacksmithing instruction with a 70-year-old teacher. Their day started at six in the morning and lasted for nine hours. Before their work experience began in earnest, each boy made his own anvil, forge, hammer, tongs and other small tools. Half of their time was spent at a shop, shoeing horses; the other half was at a coal mine shoeing ponies and sharpening tools.

Following his apprenticeship, George began work with the Midland Scottish Railway Co. in London, where 1,700 Shire delivery horses had to be fitted with shoes before being hitched up to pull the streetcars.

The Ainsworth blacksmith shop in Halkirk was a bustling enterprise. Settlers were flooding into the area and there was a constant need for farrier's services or someone to fix a wagon wheel or sharpen a plough. Large farms with livestock were being established, and the grain industry was steadily growing. The first business to open in the new village of Halkirk was a small restaurant, quickly followed by a general store moved in from the nearby Higgins stopping house, and a hardware and furniture store.

David's oldest brother, John, with his parents in the family's Model T Ford.

Owing to the area's agricultural base, a flour and feed store was established next, along with a John Deere Machinery agency.

George Ainsworth had a reputation for doing things his own way, and he was a bit of a perfectionist. He charged more than other farriers and blacksmiths, but his work was considered superior.

"I remember him putting his arm around my shoulder one day — I was probably 9 or 10," says David. "He said, 'Only two blacksmiths ever went to hell: one of them tried to work with cold iron, and the other one didn't charge enough.'"

As old-timers in the area would recount after his funeral in 1958, when George Ainsworth put a shoe on a horse, it stayed on. If a horse was developing a limp or certain signs of crippling, he could sometimes fashion a shoe that would compensate for the problem. He was good with horses but he didn't have a lot of patience with them if they were poorly trained. And if he had to shoe a particularly unruly animal, there were undoubtedly a few choice words for the owner.

With a thriving business and their first year of village life drawing to a close, George and Ellen had the distinction of providing Halkirk with its first newborn citizen. John Halkirk Ainsworth was born at home with the aid of a midwife on December 26, 1909. He was the first of the couple's three boys.

John's middle name was a fitting tribute to the newly incorporated town, but if it was up to his father, the boy never would have been given a middle name.

George Ainsworth was determined to lose any hint of "Englishness" he might have brought with him across the Atlantic, since "green Englishmen" were the laughingstock of any prairie community. Young and spoiled, with little common sense and even less ambition, their reputation was so bad that it wasn't unusual to see signs on businesses proclaiming,

Despite poor crops, George Ainsworth used to joke that "at least I got my seed back."

"Englishmen Need Not Apply." They were an embarrassment to their countrymen, and George did everything he could to blend into his new surroundings — including learning to speak without an English accent.

When it came time to name his first-born son, he didn't want any part of the British tradition of bestowing several middle names upon the child. In fact, he didn't want to give John a middle name at all. The women of Halkirk were quick to voice their objections, however, and with Ellen on their side, they overruled George's edict of no middle name.

When the couple's second son was born six years later on October 2, 1915, George and his anti-English campaign prevailed. This boy would be called Tom Ainsworth. Not short for Thomas or anything like it. Just plain Tom. And no middle name.

By the time David Ainsworth was born six years after Tom, George relented slightly by allowing the more formal "David" rather than insisting on "Dave." But that was as far as he'd go; like Tom before him, David would have no middle name.

When considering the personalities of the boys' parents, Ellen was the more sociable of the two. She enjoyed sharing family stories, and when time permitted, participating in community groups such as the local school board and women's auxiliary.

George Ainsworth was on the quieter side. He liked nothing better than to smoke his pipe and read a newspaper in his spare time. In matters of child rearing, he was strict but fair. There was never really a reason for him to raise a hand, recounts David. "We got the idea without him actually taking action."

As Halkirk grew, the pace of work at George's blacksmith shop quickened. The village now boasted 200 residents, and coal had been discovered north of the townsite stretching to the banks of the Battle River. Before long there were five producing mines within a five-mile radius of Halkirk.

Seeing an opportunity for business expansion, George Ainsworth became the local agent for the McCormick Deering Company, which became International Harvester. He later rented out part of his shop for what became the first garage in Halkirk. However, the long hours he was putting in were taking their toll.

George's blacksmith shop kept him busy from sunrise to well into the evening nearly every day of the week, despite the fact that he had two men working for him. It was all he could do to keep up. Meanwhile, although the family had been village residents for the past 10 years, he hadn't abandoned his original dream of owning a farm. He was watching immigrants from Europe who had a background in farming take up homesteads in the area and make a successful go of it.

David, aged 14.

Before long, the grass was literally looking greener on the other side. In February of 1920, George decided to sell his blacksmith business and try his hand again at farming, this time purchasing an entire section of land, 640 acres, just northeast of Halkirk. Although the blacksmith shop and farm equipment agency had driven him to exhaustion, they had been profitable and had enabled him to purchase the farmland outright.

He completed the task of having the property fenced, then set about drilling wells. He soon discovered there was plenty of water, but due to the layers of coal throughout the area, much of it was as black as ink. This lack of a decent water supply proved to be a major drawback throughout the next two and a half decades, making their hardscrabble existence that much tougher.

That first summer on the new farm, Ellen Ainsworth gave birth to their third son, David, on June 13, 1921. John and Tom were big babies at birth and David was no exception. David's days as a toddler were spent largely with his mother, almost as an only child. Tom was off at school during the day, and John was now finished school and working on the farm.

Like many children who grew up in rural Canada during the 1920s, David attended a one-room school that housed 15 or 20 students from Grades 1 through 8. The area around Halkirk was dotted with as many as a dozen such schools. They would open and close from year to year depending on enrollment figures and whether the parents could raise sufficient funds to hire a teacher.

"I remember when I went to school, there were kids there who were 15 years old and they were in Grade 2 or 3," says David. "Because of the circumstances, there just wasn't a school there before, so some families had to keep their children at home to work on the farm or get home schooling until there were enough people in the area to start a school. Other times you could have four or five kids in one family, different ages, and they'd all start school together in the same grade."

Fortunately for David, Island School #4013 had been open for several years and was located close enough for him to attend. On his first day of classes at the age of seven, he climbed up on a horse behind his brother Tom, lunch kit and notebook in hand, and trotted off down the road to begin his education.

Growing up on the homestead. From left: John and David Ainsworth, with a group of friends.

He was an interested student and on more than one occasion managed to complete two grades in one year. With eight or nine grades being taught in one room, this wasn't difficult to arrange.

Combining two years into one also permitted young David to get his Grade 9 at the school, which was a rare occurrence. Officially, the school went only to Grade 8. But the teacher, despite being in a fog as a result of his fondness for the bottle, was able and willing to steward David and one other student through the Grade 9 curriculum. (Some 15 years earlier, David's oldest brother John had to take Grade 8 twice since no instruction for Grade 9 was available, and he was too young to quit.)

Although he certainly had the aptitude, continuing on with a high school education was impossible for David. The closest high school was in the next town, which would have meant boarding in Castor or having daily transportation there and back. More importantly, this was the beginning of the Depression, and moving off the family farm for high school, even if it was only six miles away, was a luxury that many prairie families could ill afford. John had already left home in search of work, and Tom would soon follow. Someone had to stay and help George look after the farm.

Even without attending high school, David was already getting an important education on the farm, especially in the trades.

In addition to all the mechanical and repair work that goes with farming — especially in lean times when self-sufficiency is crucial to survival — George Ainsworth would occasionally do blacksmith work for neighbours and former town customers to supplement his income. When he had sold his shop in 1920, he kept his forge and various other tools of the trade.

"I worked by my father's side a great deal," says David. "When you were going to school, even though you were only seven years old — which sounds kind of funny now — you didn't go out and play with the kids when you got home if there was work to do. If it was a big job, we'd be doing the work on Saturday too.

"Tom was a big kid, too big for his age really, and he was expected to do things lots of times that I came to think later weren't exactly fair. A kid of 12 or 13 is going to dawdle around a little, yet he was expected to have the responsibilities of a man because he could pick up the big hammer."

When red-hot steel came out of the forge, Tom would follow George's rhythm of pounding the steel on an anvil: George's lead hammer making a "bing"; Tom's short-handled sledge hammer following quickly with a "bang." *Bing-bang, bing-bang* they'd go until the steel cooled and had to be reheated in the forge. David's turn at the anvil came as soon as he was big enough to wield the heavy hammer and keep up to his father's pace.

He didn't know it at the time, but the groundwork was being laid for a lifelong affinity for tools and machines and engines, not to mention a host of other skills that would prove invaluable over the coming years. Seeds were being planted in other ways, as well: David's father had a prodigious work ethic, and he was a perfectionist.

> *The Halkirk Literary and Debating Society held its meeting in Higgins Hall. Mr. G. Ainsworth took the chair. The debate that evening was "Resolved That Drink Causes More Death and Misery Than War and Pestilence." The judges were in favour of the negative.*
>
> — HISTORICAL ACCOUNT, FEBRUARY 23, 1910

The stock market crash of 1929 caused worldwide economic problems, and rural Alberta was hit hard. Farmers were forced to sell their livestock for a pittance in order to qualify for relief food for their families and remaining animals. Businesses collapsed in the once-thriving towns. As if the mass unemployment and poverty wasn't bad enough, the 1930s on the prairies were marked by severe droughts, dust storms and plagues of grasshoppers. Getting a meager two bushels of wheat to the acre, George Ainsworth would joke dryly that "at least I got my seed back." Even that was rarely the case.

For teenagers growing up around Halkirk in the 1930s, the choices for entertainment were limited. Cars were rare, and many of the roads were little more than dirt paths before they were upgraded when gravel was eventually hauled in by train. Even a good saddle was a luxury, so the kids usually rode bareback. A softball game at a neighbouring school was as good a reason as any for traversing the fields and coulees to meet up with friends.

Driving a wagon with a team of six or eight horses also brought its own form of farm-boy humour. David and a buddy would be sitting in the wagon and the friend would ask, "What are you doing these days?" The standard reply was, "Looking at a bunch of assholes." That got them giggling every time.

Around this time David's group of friends grew to include another 16-year-old, Susan Rowland, or "Susie" as she was known then. Her parents' farm was just a few miles from the Ainsworths' — their mothers had been friends for years — but she had attended a different school, right through to Grade 8.

Susan was the second oldest of 12 children — 5 boys and 7 girls. Her father, Thomas Rowland, also came from a family of 12. He was born and raised on Manitoulin Island in Ontario and was known for his skills as a horse logger.

Susan Ainsworth's father, Tom Rowland, was a horse logger in Ontario before moving west in 1917.

Tom Rowland (fifth from left) with a woodcutting crew in Ontario.

"My father came to Alberta in 1917," says Susan. "I think part of Stanley Park in Vancouver was going to be logged and my dad and his brother-in-law loaded onto the train with his horses for the trip from Ontario. It was wintertime when they left, and you could only go so long on the train without unloading all the horses, feeding them and moving them around — they didn't realize this before they started. By the time they got to Calgary, that was as far as they wanted to go. They had the Rocky Mountains coming up and they didn't feel the horses could stand any more, so they stopped in Calgary because he had a sister living in Delburne."

Soon after their arrival, Tom Rowland met a woman named Sadie Dobson. She was an only child whose family had moved from New Brunswick to Calgary in 1906 and on to a homestead near Stettler a year later.

The couple married in 1918 at the home of Sadie's parents, rather than a more public venue like a church because of the worldwide Spanish flu epidemic then raging across the west. Sadie was 21, Tom was 33, and within four years they had their own farm. It was around this time their daughter Susan was born, on January 11, 1922.

With cattle, horses, pigs and other livestock to care for, many of the daily chores fell to young Susan and her siblings, often with Susan in charge of the younger ones. Whether it was canning endless quarts of wild strawberries or digging potatoes from the family's large garden, everyone was expected to pitch in.

As the Depression wore on and families struggled to get by, relief food was delivered to area farms. Aid packages contained everything from apples and cheese to canned herring and flat pieces of dried fish, which were dubbed "snowshoes." As a matter of pride, Tom Rowland

Susan Ainsworth, standing at back, was the second oldest of 12 children.

insisted he could feed his family and refused to accept the packages, but they were left at the house nonetheless.

While the large Rowland family never went hungry, the abundance of children did put a strain on the pocketbook. Their grandmother made them clothes from flour sacks, and castaway coats were cut down and fashioned into new ones. If Susan or her siblings complained, their mother had a ready reply: she would tell them that when she was growing up as an only child, she wanted for nothing yet was still unhappy because she had no one to play with. "You've got brothers and sisters," she'd say, "so never mind your shoes."

Trips into town were a rare treat. One of the highlights of the year was attending the Halkirk annual sports day on July 1, when Sadie Rowland would take her troop into town, lunches packed, and splurge on a jug of lemonade from the Chinese restaurant to round out their holiday picnic.

Tom Rowland didn't like his children off playing at other people's homes, so friends usually came to their house. The Rowland farm was a veritable playground where everyone enjoyed kick the can, hide-and-seek, grey wolf and baseball in the summer, and ice skating and sleigh rides in the winter. One friend of the family, Art McKim, recalled playing horseshoes so far into the evening that they had to put white paper on the pegs so they could still see them in the fading light.

> *One day the guy who was paying me at that job near MacLeod came around and he had a new Ford car, which was rather rare. Ford and GM and everybody else were still building trucks for the army. This car wasn't all that fancy but it was pretty fancy to me because it was new. He wanted to take me out for a ride, so we got onto the highway where there was paved road. There were three of us in the car, and this guy just drove it wide open the whole time. This was pretty fast for someone who had never been in a new car, and it was sure a novelty for me. He drove the hell out of that car though. If I had had that car, I would have taken such good care of it that I never would have wanted to touch it.*
>
> — DAVID AINSWORTH

When a young David Ainsworth started coming around to visit the Rowland household, he was well liked by all the children. More impressive yet, especially to Susan, was the fact that he was able to drive his father's Model T Ford. It was an older model acquired by George Ainsworth back in the days when he sold farm equipment. David had managed to squeeze a few more years out of it by rebuilding the engine. Now their group of friends had transportation, even if the occasional drive to Stettler for a movie was all they could afford. On one notable trip to a school dance, there were 14 excited kids hanging off David's Model T.

During winter months when the local roads were impassable, David would walk the two and a half miles across the snow-covered coulees and pastures to visit the Rowland house. By and by, he and Susan were spending more time together, and the couple eventually married on December 9, 1940, when they were both 19 years of age.

Canada had by now declared war on Nazi Germany, but the effects of the Depression and drought were still being felt more than the war throughout southern Alberta.

David started out his married life working on a neighbour's farm, but soon travelled to Calgary in search of better-paying employment. After hiring on as an electrician's helper, he was eventually sent to a large, windswept field, in the middle of which sat a small mountain of pipe. The 20-foot lengths were to be used for a building at the Fort Macleod airport, which was under construction. His job was to cut a 12-foot piece off each pipe and then re-thread it, all by hand.

The only building in sight was an outhouse, and the man who hired him seldom paid a visit. It was tedious work, but he was making $1.50 an hour, all the more important since he and Susan now had an infant son, Allen, who was born on June 19, 1941.

David and Susan Ainsworth were married in Stettler, Alberta.

Six weeks of toiling in solitude on the open prairie seemed like an eternity, but his pipe-threading days were about to end. David's older brother Tom, who had left the family farm for the lush forests of Vancouver Island, sent word with an offer of work. Tom was now living at a logging camp on Cowichan Lake, operating some of the first chainsaws ever to be used in the region. His crew needed a "stooge" right away, someone to carry the gas cans, wedges and tools used to keep the saws running.

Upon getting word from Tom, David recalls that he "dropped everything like a hot potato" and got on his way out to B.C.

Chapter 2

CORK BOOTS
AND CHAINSAWS

Industrial Timber Mills' Camp 3 at the west end of Cowichan Lake was home to 300 men.

By 1940 the Cowichan Valley area on Vancouver Island already had a long history of logging activity, complete with its own boom and bust cycles. It was one of the most rapidly developing regions on the Island during the 1920s. Steam-powered ground-lead yarders had replaced oxen and horses for moving logs, and the addition of two main railway lines — the E&N and CNR — fostered increased logging activity by large and small operators alike.

Lakeside logging camps flourished, and soon another round of buyouts and new enterprises began. Just before the crash of 1929, a newly formed company named Industrial Timber Mills Ltd. (ITM) opened a sawmill at the townsite of Youbou (a place name combining the surnames of C.C. Yount and G.D. Bouton, both operators of the Empire Logging Company).

Unfortunately, as soon as the mill started up, the bottom fell out of the world economy, signalling the end of this particular logging boom in the Cowichan Valley.

By 1933, only 10 of the original 19 companies in the area remained operating. By 1934, however, timber prices and markets had improved and ITM was in a position to grow.

ITM's Camp 6 on Cowichan Lake, which had closed in 1931, reopened in 1934, and the company took over a second camp at the west end of Cowichan Lake, known as Camp 3. With both camps in full operation, log production in the area doubled by 1936 and remained steady through the next decade.

It was at Camp 3 that David Ainsworth's brother Tom found employment after a logging stint at Franklin River west of Port Alberni. The Second World War had begun the previous year and lumber was in high demand for everything from shipbuilding to barracks construction. At the same time, the war effort meant that few qualified workers were available for the B.C. forests.

Camp 3 was home to 300 men aged 20 to 55 with colourful job titles like bull bucker, donkey puncher, whistle punk, chokerman, flunkey and rigging slinger. They were an independent lot who took great pride in their jobs and the skills involved.

David Ainsworth (far left) with a logging crew after felling a huge Douglas fir. A tree of this size could yield up to 50,000 board feet of lumber.

Working in the woods was difficult and dangerous, and the conditions on the Island were often wet and miserable.

This was a period of immense change in the industry. Steady advances in logging systems, particularly in how logs were yarded through the bush and transported to the mills, contributed to major increases in production, especially for bigger companies that could afford the new equipment.

In the meantime, the valley bottoms with the largest-diameter trees were gradually being logged out. Next in line were the nearby hillsides, where the average tree diameters were three to five feet rather than the six to eight feet or more found in the lusher valley bottoms.

The age-old method of falling and bucking trees by hand with axes and cross-cut saws was no longer keeping pace with the rest of the industry, and the search was on for a practical means of mechanized falling. Experimental versions of the modern-day chainsaw date back

The steam-driven locomotives rattled their way right through the centre of Camp 3.

to the early 1920s, but it wasn't until 1937 that the two-man gas-powered chainsaw made significant inroads into West Coast logging operations.

It just so happened that big prairie farm boys were ideal candidates to operate these noisy new contraptions. The saws weighed up to 210 pounds when filled with gas and oil, and they had bars ranging from four to seven feet in length. They also required constant maintenance. Young bucks like Tom and David Ainsworth were well-suited to lug these behemoths through the bush all day, and they had the necessary mechanical skills to keep them running. It was extremely demanding work and the positions were difficult to fill — so much so that fallers, as well as men in other essential service positions such as shipbuilders, were exempted from the draft and "frozen" on the job.

The tasks of hand-falling and bucking traditionally went to immigrants from Scandinavia and central Europe, although their numbers were on the wane by the mid-1930s. They worked harder than anyone else and had a long history of using axes and cross-cut saws — also known as Swedish fiddles — to fell timber. Bringing down an eight-foot-diameter Douglas fir with a cross-cut saw and axe while balancing on a narrow springboard over a steep side hill required nerves of steel, tremendous strength and endurance, and no small amount of expertise.

All of this changed with the advent of the power saw. The early models were two-man machines, which meant the head faller at one end would guide the saw's long bar through the tree. At the machine-end of the saw, the second faller or machine man would actually operate the saw, keeping it running and hauling it to the next tree to be cut. They called this job "working in the engine room."

Many of the Swedes and other hand-fallers became head fallers on the power saw crews, but very few became machine men. They didn't like power saws to start with, and few were mechanically inclined. For men accustomed to day after day in the bush with nothing more than the rhythmic sound of a cross-cut saw or finely honed axe chipping into a tree, these power saws were impossibly noisy, awkward and heavy. Not only that, the hand-fallers' numbers were dwindling, and there was no one stepping in to pick up the axe.

Right: Logging a Douglas fir, David Ainsworth is on the machine-end of a Reed Prentice saw, with head faller John Svedt. This 1942 photo was taken to promote the sale of wartime Victory Bonds.

This is where prairie farm boys came into the picture.

"I went over there to be a stooge on one of the chainsaw crews; they were only running about three chainsaw crews at the time," says David Ainsworth. "I had to buy new cork boots and everything. And I remember being out in the woods in this big timber and thinking that I didn't know a fir from a hemlock, but I had to catch on pretty quick. Of course, the stooge carried the gas cans and wedges and stuff like that, and once in a while they'd give you a chance to get in on the saw. Tom was anxious to see me get started and the bull bucker or foreman was a man by the name of Stan Mahoney. He liked these new chainsaws and wanted to get more men trained to use them.

"They were continually bringing people out hoping to find someone who could run these things. They'd send a guy into the woods, he'd take a hold of the handlebars and wouldn't even be able to lift it off the ground. He'd feel how heavy it was and say, 'No thanks, I'll go in the army. I'm not carrying that thing around on a side hill.'"

The first chainsaw models used at this stage by Camp 3 crews came to be known as "Hitler" saws once the war started because they were manufactured at the Andreas Stihl factory in Germany. Breakdowns were the order of the day, and prior to the war, replacement parts had to be shipped from Germany. When Stihl's patent protection ended with the start

By 1943 trucks were being used to haul logs in the Cowichan Lake area.

Left: Tom Ainsworth on the machine-end of a seven-foot chainsaw blade. A Douglas fir like this would be 200 feet tall, yielding five 40-foot logs.

Tom Ainsworth (right) could fall a tree this size in half an hour if no notches were required.

The power saws could also be used for bucking the tree into lengths. Adjustable collars permitted the blade to be turned upright after cutting left or right.

Spar poles up to 188 feet high anchored elaborate cable systems used to yard the trees out for loading.

of the war, numerous companies in B.C. and farther afield immediately began to manufacture their own models, largely based on the Stihl's design. Within a few years, Industrial Engineering Ltd. became the manufacturer of choice for its two-man and later one-man models.

The advantage of the power saws over hand-falling became apparent almost immediately. Even with the problem-plagued early-model chainsaws, a crew could cut anywhere between 20,000 and 50,000 board feet a day. Hand-fallers managed about 7,000 to 13,000 board feet a day. When lumber companies were forced to get into smaller-diameter timber, hand-falling became even less productive.

Early chainsaws weren't well-suited to bucking trees — cutting a long log into shorter lengths — but they could be used for this work. In most cases, a single tree would yield five 40-foot logs. A float-type carburetor made it difficult to use the earlier models upright, and if the bar with the chain got pinched, it had to be disassembled to be removed.

Falling crews got paid by the amount of timber they cut, roughly $1 for every 1,000 feet. They called it "bushel work." Cutting an average of 20,000 to 35,000 feet of timber daily earned each logger $10 to $12 a day.

"It was up to the machine man to start the saw and keep it running," says David. "The motors seized up when they got clogged with sawdust, and broken chains had to be fixed on a stump. Sometimes you'd have to hold your hard hat over the damned engine while you were cleaning the points or something, just to keep the thing out of the rain."

Just getting around the bush was an ordeal. Crews would often have to traverse ravines and gulleys with a downed tree as a bridge — usually a two- or three-foot-diameter "sapling" — packing all their saws and gear with them. While veteran loggers took such crossings in stride, the rookies in the camp — David included — had their share of tumbles.

In bushier areas, the woods were often overgrown with the dreaded devil's club, a prickly, large-leafed plant that grew on a thick vine.

"Devil's club was the worst thing in the woods," says David. "The plants had spikes on them up to an inch long. Sometimes you'd have to clear them away with an axe in the really brushy areas. Other times, the head faller might be headed through a patch of devil's club and he'd tramp some down or bend some as he was walking. Then when you came along with both hands full carrying a saw or some equipment, these vines would flip up and hit you right in the face. Well, that's too bad, you just had to put up with it."

In the Camp 3 environs, the men were usually working on side hills, and fallers and buckers were each given a portion for cutting. They'd start from the bottom of the hill and work back and forth, laying the trees out on an angle. The head faller or bull bucker would pick the tree and decide which direction it should fall.

If the tree's diameter exceeded the length of the saw, the faller first had to cut side notches into the trunk so the head of the saw would fit in. Next the faller would chop out a series of horizontal undercuts with a specially designed axe featuring a sharp pick on one end. The faller had to make these horizontal cuts rather than the V cuts used in hand-falling because chains at the time couldn't cut across the grain. In the final step, the faller did falling cuts or back-cuts, which made the tree fall in the chosen direction, usually sideways along the hill. If the tree was particularly big, crews would use a hand-saw to finish the job when the bar wouldn't reach. To preserve the big timber's premium value, they'd often drop three or four smaller trees as a bed to prevent any breakage or damage.

> *When Tom [Ainsworth] was a faller at Camp 3, he slipped on some loose bark and fell and broke his leg, so they had to get him to Youbou. They had to pack him out of the bush, which took about an hour, then another hour or so on the speeder. My mother was waiting on the station platform there. When they got him to Youbou the superintendent told my mother: "You've got quite a man here; he's been swearing steady for two hours and he hasn't repeated himself once."*
>
> — LEE AINSWORTH (TOM AINSWORTH'S SON)

Experienced crews with a skillful head faller could accurately drop an eight-foot-diameter fir or cedar right where they wanted it, but there were still many things that could go wrong. If the head faller gave that seven-foot bar a shake from his end, the machine-man would immediately look up for limbs and branches falling from above. Although fallers for big outfits like Industrial Timber Mills were among the first to be required to wear hard hats, these offered little protection from a 100-pound branch crashing down. Such limbs were known as "widow makers."

"When we were falling the larger timber, the machine man would turn the saw off at the first sign of danger and get ready to run to a spot that we'd already picked out, maybe even cleared beforehand with an axe," says David. "If it was a dangerous tree and something was going to go, there was no time to think. We used to say that loggers needed feet like wings."

The same procedure was used when the necessary cuts had been made and a tree was ready to topple. Once it started to fall, the loggers would drop the saw and run for cover. Trees with a rotten middle or leaning too far could easily split or throw off pieces that could cause serious injury. One particularly lethal version of this was the "barber chair," the name given to a heavily leaning tree that split unexpectedly, leaving a sharp, spire-like section protruding from the stump. Barber chairs were usually caused by improper facing or back-cutting.

Crew of Ingvald Mickelson, Carl Olson (with bucking saw on shoulder) and David Ainsworth felling a tree near Shaw Creek.

On the rare occasions when the Ainsworth brothers ran into a barber chair, the suddenness of the split and the deafening crack terrified them.

"These things could be 40 or 60 feet high before they'd break off and come crashing down," says David. "You'd really have to keep your eye on them because there was no way to get back far enough to get clear. They were pretty damn scary and people got killed."

Huge "stump pulls" or extruding pieces at the butt end of a log could knock down anything in their path when the timber was yarded out and transported to the mill. The required practice was to cut off the offending pieces in the bush, but a crew intent on cutting corners might let a stump pull go, potentially causing accidents farther down the line.

The fallers had more than their own lives to worry about. They also had to ensure there were no wayward trees or limbs falling on their buckers, who would often be following close behind, cutting the trees into more manageable lengths.

The hardship and dangers of the job notwithstanding, there were many days when the young coastal logger was only too happy to be at work in the woods.

"There were some wonderful times," says David. "Some of the best days we ever made on the scale were with trees that were only four feet or a bit better at the stump; they just stood there like carrots. In conditions like that I think we did 40-some trees in one day.

"While you were cutting one tree, the head faller would be looking at the next tree. We were literally running from one tree to the next as the most recently cut tree was on its way down.

"One time in small timber like this, there was a guy working on the rigging not so far away from where we were. When we got back to camp at quitting time, he said, 'I was listening to you guys, and that chainsaw sounded like the Flight of the Bumblebee.'"

The introduction of the power saw changed the logging industry in a fundamental way, but this wasn't the first time mechanization had a major impact.

Whether those trees were hand-felled or cut with a chainsaw, they had to be hauled out of the bush and transported to the rail line for the journey to the Youbou sawmill. The days of dragging trees out of the bush with horses and oxen were long gone. High-lead and aerial yarding systems were now common, driven by powerful "steam donkeys" and using steel cables, towering spar trees and elaborate rigging arrangements geared for varying terrain and conditions. Loggers had to learn a new set of skills involving hooks, blocks, chains, carriages, chokers, swivels, tongs and clevises, not to mention miles of wire cable of varying thicknesses and types. What was already a dangerous occupation became even more hazardous.

The high-lead yarding systems were introduced after the First World War, and they were controversial from the beginning. Large logs hooked onto a cable and yanked down a hillside by the steam donkey would crush everything in their path, whether it was young trees needed for regrowth or potentially merchantable timber. High-lead yarding also left a huge amount of wood debris or slash behind, which only added to the possibility of forest fires in the hot summer months.

Overhead cables could snap and drop a load of logs 100 feet to the ground, scattering everyone below who happened to be working in the vicinity. Increased mechanization also put additional pressure on crews to boost production and on supervisors to make the logging operation more efficient. In many cases, owners and foremen placed a higher priority on efficiency and speed than on the safety of the employees. In the period between 1930 and 1939, some 449 loggers were killed in the B.C. forest industry. Between 1940 and 1949, the figure rose to 591 deaths. When loggers heard seven long blasts of the steam whistle echoing through the woods, they knew another man had been killed on the job. The seven long blasts were called the "misery whistle."

The production increases afforded by these mechanized systems were too large to ignore, however, and they were dominant in the industry by the 1920s.

With concerns about job safety and disagreements over wages, loggers made various attempts to form worker associations and unions over the years, against considerable opposition from employers. They had varying degrees of success until the International Woodworkers of America or IWA emerged as the dominant representative body in 1944.

"My brother and I were among the first ones who really got active in trying to get the union organized there," says David. "We needed a union, for safety as much as anything else. We made a little progress, but no sooner had we got the union sort of going and struggling, than people moved in and tried to take over."

In spite of job conditions at Camp 3 that could still stand improvement, Industrial Timber Mills probably did as much to promote safe work practices as any company at the time. In fact, the safety training in the bush that David received there would serve him well when he set out on his own.

Tom Ainsworth on the front of a steam-powered locomotive at Camp 3.

Once the logs were yarded out of the bush to a landing, they were loaded onto railcars pulled by a steam locomotive, or "locies" as they were called. The U.S.-made Climax models used on the steep terrain around Cowichan Lake were "geared locies" rather than the "rod engines" used on more conventional rail lines. The rod engines lacked the power to pull heavy loads up steep grades and were too large to negotiate sharp curves. The geared locies featured a driveshaft that was connected to each axle, providing significantly more power at much slower speeds — at grades of up to 12 per cent.

David and Tom Ainsworth in Vancouver around 1941, en route to Alberta for a visit.

Their smaller size also made them more suitable for the backwoods tracks, although they were known to have the roughest ride among several similar makes. They weren't allowed to exceed 10 miles per hour pulling a load on a downhill grade, which probably wasn't a bad idea given the area's steep terrain. Locie engineers and their crew frequently had to deal with broken and spread rails, rotten ties, washouts and rockslides, weakened bridges and trestles, oncoming trains or speeders, overloaded cars and mechanical failure.

Not surprisingly, there was no shortage of derailments. Crews had various methods of getting a loaded railcar back on track, but an overturned engine could mean shutting down the camp until the tracks were cleared.

The quality of coastal logging camps in 1940 varied greatly, but with 300 men on site and qualified labour in short supply, Industrial Timber Mills ran a tight operation at Camp 3. Every day would see a crew working, a crew leaving and a crew arriving.

"We would be about eight men to a bunkhouse, and when the lights flickered or dimmed just before 10 p.m., it meant five minutes until lights out," says David. "The bunkhouses were all in a row and the railway track went right behind them. They'd have people on the locomotive all night keeping the fire and steam up so it would be ready to go in the morning, so that thing would hiss and bump all night. Then, at a certain time in the morning — some ungodly hour — that guy on the locomotive would blow the whistle. If you were still in the bunkhouse, that whistle would just about blow you out of bed. They had their methods, and some of them worked."

Nobody went hungry at Camp 3, but feeding 300 famished loggers in one sitting required procedures that wouldn't be out of place in a boarding school. The waiters, or "flunkies," would ring a steel triangle, otherwise known as a "gut-hammer," and the men would come running.

> *Why is it that your Climax engine has to ding and dong and fizz and spit and bump and chug and hiss and pant and grate and grind and hoot and toot and whistle and wheeze and jar and jerk and howl and snarl and puff and groan and thump and boom and smash and jolt and screech and snort and slam and throb and roar and rattle and yell and smoke and shriek like hell all night long?*
>
> — Author Unknown

Each man was assigned a permanent seat and new arrivals to the camp had to wait at the door until everyone else was seated. Then the flunkies would tell them where to sit.

It only took David a meal or two to realize that he was going to starve if he didn't learn to eat faster.

"All my life — up until this point anyway — I was rather slow at eating. Lots of times at home, everybody else at the table would be done and I'd still be munching away. I'd get in trouble with my brothers because they wanted dessert to be served. You could do this in a family atmosphere, but not at camp. I learned pretty quick that these flunkies would start picking up the food before I got my second helping. And when you work that hard, you sure have your eye on getting another helping. Some of these loggers ate like wolves anyhow, so it wasn't a problem for them, but it was for me. I sure learned that I had to get on with it."

The camp's dining hall was no place for idle chatter either. Except for the occasional quip to a friend or a "please pass the butter," general conversation was discouraged.

Before breakfast in the mornings, David would join the other men in the lunch room where they made their sandwiches and filled their thermoses. They'd have just enough time to take their lunches back to the bunkhouse before the flunkies would ring the bell for breakfast.

Because Camp 3 was what they called a "railway show," the men went into the bush each day on a "crummy," the term for a special railcar or truck that transported workers to the job site. Their chainsaws, gas cans and tools would remain outside up at the front of the car, and the men would sit on benches until the train rolled by their appointed stop.

The locomotives travelled at an average of 15 to 20 miles per hour, so the trip could take anywhere from 30 minutes to two hours, depending on the whereabouts of the logging site. After a full day in the bush it was back to camp on the crummy, with just enough time for a wash and a quick nap before the dinner bell rang.

Susan and Allen near family cabin at Camp 3.

David greets son Allen after another day of logging in the Camp 3 environs.

Evenings at Camp 3 were decidedly low-key. Some of the men indulged in a bit of gambling, but there was zero tolerance for any alcohol or fighting. This was the routine for six days a week. Sundays were usually set aside for washing clothes and doing other personal chores, as well as ball games against neighbouring camps, or hunting and fishing.

Because the nation was at war, David and Tom Ainsworth and other camp men signed up for after-hours training with the Pacific Coast Militia Rangers. If Japanese invaders ever landed a force on the west coast of Vancouver Island and tried to make their way inland, they would be met by civilian militiamen who were at home in the woods and could use a firearm. Unfortunately for David, this was in the latter days of his Camp 3 experience and he had moved on before his militia-issue rifle arrived, something he's always regretted.

Camp 3 was home to several hundred men, but it also offered cabins for families at a separate site a few miles east, near the head of Cowichan Lake. As soon as she and David could manage the arrangements, Susan Ainsworth made the 24-hour train trip from Calgary to Vancouver and then on to Vancouver Island, with infant son Allen in tow. Susan's arrival at the camp for married couples was aboard a gas-powered crummy, a windowless boxcar with seats up the middle and along the sides.

"Another logger had just quit and left the area with his family, so we were able to buy his cabin for $300," says David. "They left a few basic pieces of furniture behind, but any other household items would have to be purchased in Duncan and brought out on the crummy."

David and Susan's friends from Alberta, Jack and Doris Gordon, also made the trip to Camp 3, but within a short time returned to Halkirk, much to Susan's disappointment. Doris was a close friend of Susan's and had accompanied her on that initial journey to Vancouver Island.

Rustic as it was, this shorefront cabin was David and Susan's home for several years. Allen, meanwhile, was no longer an only child, with the birth of Brian Ainsworth on January 23, 1943.

It was a dispute over wages that brought David's time at Camp 3 to an end.

"We worked at Camp 3 until we had a bad fire in the summer of 1944. Our chainsaw crew was out falling snags after the fire had gone through," says David. "The ashes were so deep they just about went over the top of your cork boots. A dying snag might be shooting flames and bits and pieces off the top that could be picked up by the wind and carried off, so they had to come down because they were a fire hazard.

"They had hired a bunch of other chainsaws to come in and help fell these things, and I soon learned they were getting paid more than we were, because we were a company crew. I told my bull bucker but he ignored it or couldn't find anyone to talk to about it. I was still young and the war was just about over, so I quit."

With plenty of chainsaw experience under his belt, David hired on with Lake Logging Ltd., which had a truck logging operation farther down Cowichan Lake, and the family moved into a nearby auto court. Tom Ainsworth stayed at Camp 3, where he now had a job repairing chainsaws in the company shop.

David spent the better part of a year working for Lake Logging until he and Tom decided to buy their own chainsaws, hire a crew and strike out on their own as independent contractors. The war was over, they were no longer "frozen" on the job, and opportunity was knocking.

Their first contract was at a remote water camp run by the legendary "Hoot" Gibson of W.F. Gibson and Sons Ltd. at Chamiss Bay, a small floating outpost tucked into Kyuquot Sound on the northwest coast of the Island. (Hoot Gibson was known for his booming voice.)

Susan, meanwhile, returned to Alberta with their two sons and stayed with her parents and David's family. Her return to friends and relatives at Halkirk provided a welcome respite from raising two young boys within the confines of an isolated logging camp.

Cut off from civilization as the loggers were, the highlight of Chamiss Bay camp life for David was the arrival of a steamer bringing in supplies.

"These steamers went up and down the coast, and one came in every week or so," says David. "It didn't matter if it was the middle of the night or the middle of the day. When they came in and blew their whistle, everybody rushed down to the wharf and went out on that boat to buy a magazine or some chocolate bars. This was our only contact with civilization."

The logging work was somewhat different than David had encountered previously. For one thing, they made the daily trip to the logging site on a small tugboat rather than a railway crummy.

The forest grew right down to the water's edge, so the trees were felled or yarded into the water. Then the fallers would work their way up the hill as they had at Camp 3.

The plan was to drop the logs sideways on the hill, which facilitated bucking and reduced the likelihood of breaking the log. There were times, however, when a log would get away from them or a top would break off and immediately start sliding down the steep incline.

"Sometimes we'd be quite a ways up the hill," recounts David, "maybe 800 or 900 feet, and this tree would turn around and head for the water. It would go down this hill at such a velocity, probably 40 miles an hour by the time it hit the water. When this happened it was kind of spectacular, so we'd shut off the saw and just watch it go down. It would go straight into the water and just disappear. You could count the seconds, and then 40 or 60 feet of that tree would come bursting back up out of the water, right where it went in."

It was lonely and remote, but their logging venture at Chamiss Bay was proving successful. The activity came to an abrupt halt, however, with the onset of a strike by the IWA. When the camp finally shut down and the other loggers began to clear out as fast as they could, many on float planes, David and Tom loaded their chainsaws and logging gear on a coastal steamer bound for Port Alberni. The old steamer wasn't built for open water, and there were times during the voyage when David wondered if they'd ever make it.

"It was an awful trip," he says. "We were going from inlet to inlet making stops, but there were times where we'd be out there in the open water on the west coast of Vancouver Island, and these old steamers were not really built for that. There were a whole bunch of empty oil drums down below our bunks, so when this boat rolled from side to side, it rolled really bad. It was really rough and I was laying in my bunk just holding on to keep from rolling out. Remember, I was just a prairie chicken, and I kept thinking to myself, 'The next time she rolls over that far, the whole ship is going to go over.'"

The Ainsworth brothers' final destination was Cultus Lake, near the city of Chilliwack in the Fraser Valley. Some Alberta acquaintances of David and Susan had a summer cabin there for rent, and various logging operations in the nearby Harrison Lake area offered good employment prospects for experienced loggers with their own chainsaws.

Despite the dangers, working in the bush paid better than anything David would find in Alberta, so continuing along that path was a natural choice.

In the meantime, surviving the rigours of remote coastal logging for five years had provided a wealth of experience for a boy fresh off the prairie who "didn't know a fir from a hemlock" when he first set foot in Camp 3. Postwar British Columbia was now the "Land of Opportunity" for the Ainsworths, but a long road lay ahead.

Trips across the wooden trestles on the Climax locomotive were a daily occurrence at Camp 3.

Chapter 3

NORTH TO THE CARIBOO

David Ainsworth and other loggers were transported to work in this basket suspended 150 to 200 feet above the valley floor. This particular basket and "skyhook" arrangement was used in a canyon in the Ruby Creek area east of Harrison Lake. When their work site was reached, the skyhook operator would lower the basket and crew to the ground, then winch them back up at the end of the day. As many as 20 men would crowd into the 24-foot-long basket at one time. The skyhook was mainly used to yard logs — up to 40 feet long and 6 feet in diameter — to a nearby landing. The stripped-down International truck chassis travelled along thick steel cables for up to 2,500 feet. Without the skyhook for transportation, loggers were forced to traverse the canyon's steep slopes on foot, carrying all their equipment, including chainsaws and full gas cans.

The mountainsides around Harrison Lake had seen logging activity since the 1880s, but they still offered abundant stands of fir, hemlock and spruce when David and Tom Ainsworth arrived on the scene in the summer of 1945.

Experienced coastal fallers like the Ainsworth brothers had no difficulty finding work. They now owned five chainsaws and contracted out their logging services, running as many as three four-man falling and bucking crews on a single job.

The lion's share of their work came from logging companies that operated camps up and down the 37-mile-long lake. Unlike the relatively modern facilities at Camp 3 on Vancouver Island, the small logging camps around Harrison Lake were rustic outposts with little in the way of creature comforts. The buildings were often old and in need of repair, and the beds weren't much better. Loggers described these camps as "gunnysack" operations — a bit ramshackle. Any repairs that were carried out wouldn't hold up for long.

David and Tom's first camp experience in the Harrison area was at a Bear Creek operation run by local entrepreneur Earl Brett. This was Brett's first camp, and it certainly fell into the gunnysack category, but good wages and even better food made up for any hardships on the accommodation side and for the long, rough boat rides just to get there.

In addition to running a local GM dealership and his logging company, Earl Brett dabbled in some mining interests west of Lillooet and was an avid pilot who knew the area well and could always be counted on in emergencies. "I always kind of admired him," says David. "He was a good guy and was always interested in a lot of things."

A contract logger's wages — up to $30 a day in good timber — were above average for the time, but after four years of falling in the Harrison Lake area, David and Tom began thinking about a change in direction. For starters, those unwieldy two-man chainsaws were being phased out and replaced by newer, lighter one-man models, which meant more logging contractors were available for hire.

At the same time, the frustrations of keeping an experienced crew together from contract to contract began to mount. It wasn't unusual for David and Tom to have to retrieve their men from the waterfront bars in Vancouver's Gastown district after too many days of revelling between jobs.

Their main reason for a change, however, involved concerns about safety. They had both seen their share of serious accidents in the bush and had experienced close calls themselves. Tom broke his leg while logging on Vancouver Island, and David injured his back after a jarring fall in a bucking accident at Camp 3. The dangers of the job were brought even closer to home one fateful day near Harrison Lake. David was on the machine-end of the chainsaw

---

One of the companies we logged for was owned by a fellow called Orion Bowman and his son Oliver. The old man was a real odd bugger; he was a nudist who practised nudism in places where he shouldn't have.

His son Oliver was the one who did the logging, and we had made a deal with him for a job; I got along with him all right. They had their own saw and asked us if we wanted to run it, so I said we wanted $24 a day. Well, that really shook the old guy, but his son didn't complain too much because he knew he had to pay that sort of thing.

You had to go down there on a Saturday morning every second week for payday, and you had to stand around and line up as if it was something special getting a cheque. There were a bunch of us there lined up waiting for our cheques, and Orion's daughter came out of the office to give them out. She worked for someone else but also looked after the old man and headed up the office. She was really sharp; quite intelligent and quick-witted.

Well this old Orion talked with a whine, very nasal, and she kind of mimicked him sometimes. She was loyal, but she was quick-witted enough to see the funny part of a situation too. So as she *handed me my cheque, she said — all kind of whiny like Orion —"Father says you're making more money than he is."*

Well, this meant I was getting close to the end. After a few paydays it looked like the old man was going to get rid of me because he couldn't swallow that deal we made with Oliver.

We were living at Cultus Lake, so when the crummy went by in the morning I just had to walk down the street and get it. One morning I walked down there with my hard hat and lunch and boots and everything and was standing there waiting for the crummy. The crummy pulls up, Oliver gets out, walks around the back and tells me I'm fired. No phone call, no warning at all. They just tell you before you get on the crummy that you're not needed. This was the first and only time I ever got fired from a job. At the time it was actually kind of comical because it wasn't hard to get a job just then. So I trudged back to the house and told Susan I just got fired. We all had a good giggle and laugh at that; imagine getting fired for making more money than the owner.

— DAVID AINSWORTH

Picking up the new Jackson Lumber Harvester from Vancouver's Bingham and Hobbs dealership in 1950.

when a falling limb about five inches in diameter struck his partner on the head. Three days later David's foreman came by to tell him the man had died of his injuries in the hospital. It was a wake-up call for both the brothers, and they felt particularly sorry for the man's wife and family.

"We were always saying you can't ignore the statistics. Whether you were careless or not, you knew that sooner or later it was going to get you," says David. "It was a serious concern for us at the time; if we stayed in the woods we weren't going to stay in one piece. We wanted to try something that had a future in it without breaking our necks doing it."

With limited options outside of the forest industry — David mused about buying a local service station, but it wasn't for sale and he didn't have the money anyway — the brothers decided in 1950 to try their hand at operating a sawmill, albeit on a small scale. On the recommendation of Carl Reid at Vancouver's Bingham and Hobbs Equipment, David and Tom jointly purchased a Jackson Lumber Harvester portable sawmill, invented by Clinton D. Jackson and manufactured in Mondovi, Wisconsin. The first of its kind in B.C., their mill came with a price tag of $5,000.

Portable bush mills had been used in the province's Interior for several decades, but they were often primitive affairs that were set up on top of large timbers and had to be disassembled before being dragged by a Cat or tractor to the next site. The Jackson Lumber Harvester was different. It featured a steel frame on wheels and could be towed behind a pickup truck.

> *Some of the best food I ever had in my life was at some of these logging camps. That was the thing that made going to Earl Brett's, for instance, quite bearable. Earl Brett was always flying his plane, and he would come in there two or three times a week on his float plane and bring in the supplies. Ice cream was one of the things he'd bring in, buckets of ice cream, and this cook — he was a real master — would load up these big, oblong vegetable bowls, serving bowls actually. People would take so much it would make you sick just to think about it. Of course, you worked so hard in those days you could get away with it. But that was one of the big treats at Brett's.*
>
> — DAVID AINSWORTH

David Ainsworth rides the carriage on the portable sawmill at a Chilliwack-area farm.

After arriving on site, the mill could be set up and sawing timber within an hour, using a tractor's power take-off as a power supply. The Ainsworths added their own innovation of using a drive shaft from the PTO rather than the traditional belt. Not only did the mill run more efficiently, but the driveshaft set-up was safer as well.

David and Tom started out by doing custom sawing for farmers and ranchers around the Chilliwack area. They'd set their mill up on a site where the owner had already hauled his logs. Accustomed as they were to machinery and handling timber, it wasn't long before the brothers became proficient at sawing logs with their new mill.

There was a hitch, though. Farmers in the area invariably had a few six- or seven-foot-diameter logs mixed in with the rest of their timber, and the Ainsworths' mill was only designed to handle logs up to a maximum of 30 inches in diameter.

"Can you cut these big ones?" a farmer would ask. "Oh yeah, we can cut anything," would be the brothers' reply, knowing full well they'd have to improvise some way to handle this massive old-growth timber that had survived the area's previous cuts.

And improvise they did. David and Tom took full advantage of their chainsaw expertise and coupled that with recent improvements in the design of the saws themselves.

In the earlier two-man chainsaws, the chain ran around a sprocket at the end of the

Brian and Allen Ainsworth with their younger cousins, Lana and Lee, on the newly purchased portable mill in New Westminster.

Break time. From left: Tom Ainsworth, John Travers, Val Price, Irwin McGregor near Cultus Lake.

chainsaw's bar, and the other end of the bar was attached to the engine with three bolts. Removing the bar for repairs was both finicky and time-consuming.

Industrial Equipment Ltd., known as IEL, changed all that in 1948 when they introduced their one-man chainsaw called the Beaver. Rather than running around a sprocket at the end of the bar, the Beaver's chain ran around in a groove at the end of the bar. The Beaver also came with a manual clutch, float carburetor and 1.25-horsepower engine. IEL then introduced its Pioneer model, which featured a longer bar for larger west coast timber, and that was followed in 1949 by the IEL Twin, which used two Pioneer cylinders.

The IEL Twin could accommodate bars up to six feet long and was far lighter than the old two-man chainsaws. Most importantly for the Ainsworth brothers, the six-foot bar had slots in either end, which meant they could attach an IEL Twin engine to both ends of the bar. They were doubling their sawing power, and with a man on either end of the bar, they were able to cut through large logs with reasonable accuracy. Getting the right chain tension was particularly important, however. With two engines running one chain, an adjustment that was too tight or too loose usually resulted in a jam-up or break.

When faced with a large piece of timber, David and Tom would start out by levelling the log; then they'd fasten planks alongside to act as guides for the chainsaw bar. With Tom on one end of the saw bar and David on the other, each controlling an IEL Twin engine, they'd walk the length of the log, cutting horizontally as they went. As they were cutting their way through the log, they'd both slip wedges or railroad spikes into the cut to keep it open and prevent the bar from getting pinched.

"We'd have this succession of wedges and spikes stuck into the cut all down the log, and if we didn't break our chain, which we did pretty frequently, we could cut like hell with an engine on each end. That thing would eat its way down a log at a pretty good clip," says David. "And if you did this right, if everything went well, then this log would just pop open at the end and we'd take the saw out."

Next they would remove the spikes and wedges, close the cut by clamping the log together, and roll the log a quarter turn so the cut was now vertical. After using a level to ensure the log was properly positioned, they'd put the guide planks back up and repeat the procedure, cutting through the log horizontally as they went back the opposite way.

"Now we had the log cut in four quarters, so we'd skid a quarter over with a lot of brute strength — and ignorance — to the carriage of the portable mill. Then we had something we could manage with the saw. We could whittle up a very big log on that small mill."

The Ainsworths' portable sawmill could only handle logs up to 30 inches in diameter. Bigger wood had to be split first by chainsaws.

Although their two-engine technique worked reasonably well for cutting these large logs, it was a time-consuming procedure, usually carried out in the evening after their help had gone home. The occasional large log didn't present a problem, but when it got to be two or three a day, their output suffered.

When an opportunity arose to cut timber for Jones Lake Logging near Bridal Falls, David and Tom moved their portable mill in for what they hoped would be a long-term contract. The terrain was the steepest they had ever worked on, but the timber quality was above average — logs that had been left behind when crews went through decades earlier in search of the biggest trees they could find.

David and Tom's job was to saw these logs into railway ties and rough lumber, which was trucked to a mill in Hope, about 40 miles east, then graded and shipped to the United Kingdom. Unlike the railway ties and lumber used in North America, the U.K. products had to be cut to unusual widths and thicknesses. Lumber could be 2½×7, 2½×9, or "two-and-a-harf by eleven" they used to joke. The ties, which were five by ten inches and eight and a half feet long, had to have "free-of-heart centres." The U.K. customers believed that ties with the heart, or inner core, of the log enclosed were more susceptible to warping.

The Ainsworth brothers now had a crew of six including themselves, and the sawmill in Hope was taking everything they could produce.

Unfortunately, their progress was short-lived. It was the summer of 1952 and word began to spread of an impending strike by the IWA. Since the mill they shipped to in Hope was a union operation, they'd be affected just like everyone else. When the strike was finally called, the industry was effectively shut down and the Hope mill stopped taking shipments until further notice. The Ainsworths continued sawing their U.K. ties and lumber for another day

A member of the Canadian Army Engineers watches David as he sharpens the saw blade. The nearby army camp at Vedder Crossing eventually purchased a Jackson Lumber Harvester.

or two until the loading racks were full, then sent everyone home. "We went home kind of muttering that day," says David. "It tied us up thoroughly, and we were wondering what the hell we were going to do now."

After several weeks, David and Tom were looking at all their options. Carl Reid, the dealer who sold them their Jackson Lumber Harvester, had suggested more than once that their mill was better suited for the trees found in the province's Interior. "He often said to us, 'You fellas should take that mill of yours up into the Cariboo in the Interior,'" says David. "He said, 'The timber is smaller up there and more suitable for this mill. Everybody's running portable mills and moving from site to site right out in the woods.'"

With time on their hands and nothing to lose, David and Tom loaded into Tom's '38 Oldsmobile and set out for the Cariboo. In those days it was a 12-hour trip, and everything north of Cache Creek was gravel road.

They travelled as far north as Prince George, looking for an opportunity to set up their mill, but all the area's unionized sawmills were shut down and weren't taking deliveries of rough lumber. They met with the same response in Quesnel, then checked in just north of Williams Lake with a contact at Jorgenson and Wells, who logged for Western Plywood, a forerunner of Weldwood.

Western Plywood was primarily interested in logs known as "peelers" — logs of sufficient diameter, usually about 24 inches, to fit on a lathe for the production of veneer at their Quesnel and Vancouver mills. As a result, they had stockpiled perfectly good logs, some actually larger than 24 inches, which didn't qualify as peelers but could be sold as sawlogs to other mills. If the Ainsworth brothers could saw them into rough lumber, all the better.

When David and Tom went out to inspect the logs, they couldn't believe their eyes. The site was piled high with good-quality Douglas fir, enough work to keep their crew going for several months. The problem was the way the logs had been skidded and piled.

"Some haywire logger had used a D7 Cat with a blade and just pushed all these logs up into piles with the gravel and dirt mixed right in. They didn't have a goddamned idea what they were doing and they didn't care," says David. The rocks and gravel ground into the logs would make it impossible to keep a saw sharp for any length of time.

"It was such a mess. We would have loved to move up there and start cutting because they were good logs and there were lots of them, but they had been so badly piled by these Cats. Once we took a look at them, we got pretty discouraged."

The prospect of so much work sitting in front of them made the decision to turn it down that much harder. Headed south on Highway 97 with evening approaching and nothing to show for their entire trip, a dejected David and Tom made a quick decision to turn off the highway and check out Clinton Sawmills, a small planer operation located at a scenic lookout called the Chasm, about 40 miles south of 100 Mile House.

Cec Ruckle, the mill's superintendent, was talking with truckers Johnny Baker and Amos Fowler in the office when the Ainsworth brothers pulled in. When Tom explained what they were up to, and described the type of mill they operated, the trio became interested.

"They said, 'You know, you guys might be just what we're looking for. We've got a guy out there with a brand new D4 Cat who's logging west of the highway, and he's upset because he's not cutting near enough logs to keep him busy. We're trying to find a mill that would go out there and work in the same area, maybe a quarter mile apart, and he'd log for one mill in the morning and the other mill in the afternoon. You should go find him because we'd like to keep him — and keep him busy.'"

The Clinton trio gave David and Tom directions — "straight west out that road to Meadow Lake and watch for a sign that says 'A Friend.'"

The next morning the two brothers set out for Meadow Lake, which turned out to be an hour-and-a-half ordeal over a rutted, bumpy road. To make matters worse, the ruts were filled with half a foot of dust that billowed out in front of the tires like a wave of mud. Sure enough, though, they found a board with "A Friend" written on it, nailed to a tree — a poetic touch to be sure — and made their way into the work site just in time to watch a new Cat roar in, drop off a load of logs and head straight back into the bush. The portable mill operator, a man by the name of Albert Friend, just kept on working.

"The next time that Cat came in we made damn sure he understood that we wanted to talk to him," says David. The Cat operator was Stan Halcro, a recent arrival from the Okanagan, and he was all ears when he heard their proposal. He said he'd work from dawn to dark if he had to, just to keep both mills busy. This was all they needed to hear.

David and Tom went back to Clinton Sawmills and settled on a deal of $20 per thousand board feet of rough lumber. The sooner they returned with their portable sawmill, the sooner they could get up and running. Neither David nor Tom knew where this was leading but at least they had found work, even if it meant hauling their show all the way up to the backwoods of the Cariboo.

Tom and his wife Muriel, whom he had married on May 24, 1938, during his Camp 3 days, were now living at Cultus Lake and raising their three children: Diane and twins Lana and Lee.

For Susan Ainsworth, Cultus Lake was a welcome change from the remote logging camps and dreary winters on Vancouver Island. After six years in this small resort town just south of Chilliwack, she was beginning to feel at home, especially when David and Tom got out of the logging business and bought the portable mill. David was no longer away for long stretches and could even take a weekend off.

"Allen and Brian were very busy boys, and that was my whole life right then," she says.

On the Hope-Princeton highway, the Ainsworth caravan makes its way toward the Cariboo in 1952.

The "world's most portable sawmill" was pulled behind Jim McDuff's converted logging truck.

"I couldn't go to work because I never would have left my boys with anybody else before they started school. I was still used to making most everything; I knit and sewed them clothes and did all the canning and baking and those sorts of things. When you're raised on a farm, that never leaves you."

The IWA strike was still in full force, so David and Tom had little difficulty rounding up a small crew for their journey to the Cariboo. It just so happened that an acquaintance of theirs, Jim McDuff, was heading up to Clinton with his truck for a log-hauling contract. Strike-bound like so many others, McDuff had dropped his trailer and equipped his truck for hauling short logs. For the trip north, they cut some timbers and fashioned a deck on his vehicle's short-log bunks. Then they backed the rig up against an embankment and loaded on the tractor that was used to power the portable mill.

Jim McDuff also towed the portable sawmill. This gangly procession was rounded out by David in his pickup truck, pulling a board edger.

The most direct route north was through the Fraser Canyon, passing through Hope, Yale, Spuzzum, Boston Bar, Spences Bridge and on through to Cache Creek. In 1952, however, this section of road was a far cry from today's well-engineered Trans-Canada Highway.

Construction on the canyon route actually began during the gold rush days in 1860, when the Royal Engineers were asked to build a mule road to move traffic to Barkerville, some 375 miles to the north. By the early 1950s, sections of the road through the Fraser Canyon were still no better than a trail. In some stretches, which have now been replaced by tunnels, motorists had to drive on wooden shelves or benches cantilevered out from a sheer rock wall.

Many sections were only one lane wide. If two vehicles met halfway, someone would have to back up. As a result, trailers weren't permitted on the route. Freight trucks often drove the treacherous road at night so drivers could see oncoming headlights at a distance and pull over to a wide spot to give the right of way.

For the Ainsworths' little convoy, the canyon wasn't an option. The portable mill would have to be transported north via the Hope-Princeton highway. It was a longer route, but it had recently been paved for the first time and was better for trailers.

"So one morning we were all ready to go and I remember Jim McDuff's wife Betty saying, 'Jim's off on another adventure!' Everybody thought we were crazier than hell," says David, "but we were going to go to work and they were sitting around wondering when the strike was going to end."

They arrived at Clinton Sawmills late that evening. After unhooking the edger from the pickup truck, they drove back into town, where they got a room at the historic Clinton Hotel.

The next morning the first order of business was unloading the tractor from McDuff's truck. He was off on a log-hauling contract as previously arranged, so they would have to use the tractor to tow the portable mill all the way out to the work site.

The road to Meadow Lake was every bit as rough as they remembered it, and by the 10-mile mark the deep ruts were taking their toll; it was about this time the portable mill's trailer hitch broke off.

With the trip at a standstill, David drove the pickup back in to Clinton Sawmills and borrowed some welding equipment. A visit to the mill's junkyard yielded scrap metal, and then it was back out the Meadow Lake road for repairs. "I wasn't much of a welder at that time, but you have to learn in a hurry," says David. "We finally got the hitch tacked together well enough so we could get the mill the rest of the way out there. But it took us the whole day."

Once the portable mill was set up and Stan Halcro began skidding in logs, the sawing could begin. Within a few days David and Tom were joined by Jack Lefferson and Val Price, both of whom hailed from the Fraser Valley and had worked with the Ainsworths previously.

David or Tom would operate the mill and the edger while Jack and Val would pile the rough lumber into loads on sawhorses or wooden gates.

When the truck came to pick up a delivery, the driver would back into the sawhorse so the load would fall onto the truck bed. If the driver's aim was a bit off — and it wasn't an uncommon occurrence — the entire load would topple over and another round of hand piling would ensue.

As Carl Reid had told them back at the coast, their Jackson Lumber Harvester was indeed suited to sawing smaller logs, which were predominantly Douglas fir. But it took some fine-tuning nonetheless.

> *It was a real experience at the Clinton Hotel. The first night we got there the guy signed us up for a room and we had to go upstairs. The hotel was one of the oldest buildings north of San Francisco, so it had all these additions on it from every time they had needed a bit more room. They didn't have very good carpenters when they added these rooms on, though, so there might be spaces of a couple inches between the different levels of floors or the additions.*
>
> *So this guy takes us to our room with a lantern or flashlight, and every 20 feet or so you would be into another area and the floor would go up or down. Well, Jim McDuff was right behind the guy carrying the flashlight and I'm tagging along behind him carrying all my duff, so I couldn't see very good. Nobody was calling back to me to watch out for the step or anything, so every time I came to one of these spaces in the floor I bloody near tripped and fell over because it was so dark. There was probably a little bit of cursing because by now it was two in the morning and we were really tired.*
>
> *That hotel did burn down at one point, and I've always thought it was a shame we lost so many of these old historical buildings in the Cariboo.*
>
> — DAVID AINSWORTH

The Ainsworth crew, shown standing by Jim McDuff's logging truck, stayed at Ma Porter's rustic boarding house at 70 Mile for their first few weeks in the Cariboo.

"Old Albert Friend's mill had the same kind of power supply as ours, but he was cutting more lumber than us at the start," recounts David. "He was wise to the timber. We learned a few tricks after a while, though, and we would beat him every time when we had to move; his mill took a lot longer to set up."

For their first few weeks in the Cariboo, David, Tom and the crew travelled back and forth from the small community of 70 Mile House to the Meadow Lake work site each day. Their "home away from home" was a no-frills boarding house run by local legend Ma Porter and her husband Matt.

Like the Clinton Hotel down the highway, Ma Porter's place was a warren of cabins and rooms built in several stages over the decades, and the narrow upstairs hallways were treacherous. There were no baths or showers, and if a guest asked about bathing, Ma would reply that "Green Lake is down the road just around the corner."

As the Ainsworth crew soon discovered, Ma Porter's "just around the corner" turned out to be a half-hour drive along a winding dirt road, just what they needed after the long trip in from Meadow Lake.

It soon became apparent that this grinding, hour-and-a-half commute to the milling site every day wasn't working out. David and Tom needed the evenings to sharpen saws and do other maintenance work, but the long drive made it impractical to return to Meadow Lake after supper.

The only solution they could think of was staying at the site. The crew members were told to rustle up a tent on their next weekend off, and David paid a visit to the war assets store

> *We didn't run the mill on Sundays, so we'd take a day off and sometimes go swimming. You could go in Meadow Lake or some of the other lakes in the area. They either had leeches or they were alkali lakes or ponds. I liked the alkali ponds better; they burned your eyes and weren't pleasant to be in, but at least when you got out of the water you weren't hanging with all these little black things.*
>
> — BRIAN AINSWORTH

> *At Ma Porter's, Ma did all the cooking. She had this large frying pan, about two feet in diameter with a long handle on it, and she would fry up the spuds and bacon and everything right there. When breakfast was ready she would grab a broom and pound on the ceiling — we were all staying upstairs — and that would be the sign to come and get it.*
>
> *Her husband Matt was about 70, and I remember him clearly because he used to sit there and doze. When we first got there I wanted to mail a letter, so I said to Ma Porter, "I want to mail this letter." She said, "Do you see that old fart?" — she used pretty good language sometimes — "that's Matt. He's the postmaster. Go and wake him up to mail your letter."*
>
> *You had to buy a stamp from him and mail your letter, then kind of hope it would go.*
>
> *One evening Ma Porter was going out to milk a cow and Jim McDuff went along to watch. He was raised in the city and hadn't been on a farm very much. Ma Porter opens the gate and there's this big bull standing there. It looked pretty frightening to McDuff. So Ma Porter opens the gate, this damn big bull is standing there, she's wearing these high rubber boots with a milk pail in one hand and a stool in the other, and she comes up to the bull and hauls off and kicks it in the ribs and says, "Get out of here you old bugger!" McDuff just about fainted — he thought that bull was dangerous. I bet he was telling that story until the day he died.*
>
> — DAVID AINSWORTH

in Chilliwack. "I went in there and found this big army tent, big enough to stand up in, with these big flaps on the side. We started calling it the circus tent. When I got back to Meadow Lake, we set it up with a cookstove and sawdust burner in it, and we used some planks to build a platform underneath it. It wasn't too bad at all."

The commuting problem solved, the crew now needed a cook. It was summertime and school was out, so Susan Ainsworth arrived with their two boys in tow and became an immediate fixture. Being the second oldest of 12 children, she was no stranger to cooking for a group and considered her time at the camp something of a working holiday.

"I think it worked out all right," she says. "We thought it might be fun to come up there with a tent. Allen and Brian could get out in the country and I could look after David. We went down to Chilliwack every weekend — Friday night we'd go down and get home at two in the morning. Next morning I'd get up and bake like mad, buy my groceries and get back up there on Sunday night, hauling everything back up to Meadow Lake."

David now had his wife and sons at Meadow Lake to make the experience more palatable, but it was a different story for his brother Tom. Tom had suffered from asthma since his childhood days on the prairies, and the dry, dusty conditions in the Cariboo were making the camp unbearable for him.

On the financial side, the Ainsworth brothers weren't generating much of an income at Meadow Lake, and the long-term prospects didn't look promising. Returning to the logging industry at Harrison had its risks, but Tom knew he could be making far more money there than sawing rough lumber up in the Cariboo. His decision to sell his share of the business to David and return to the Fraser Valley wasn't entirely unexpected, but it was disappointing nonetheless.

"It was a blow to me at the time, no question about it," says David. "You could see it coming, though. He was anxious to get the hell out of there; I knew it wasn't working out for him and it was the best thing to do. It took some courage to return to the logging business."

Susan Ainsworth with Brian and Allen on the Fraser Canyon route.

With Susan on hand to cook and a small crew to keep busy, the thought of changing direction never entered David's mind. "Tom's leaving was kind of a tragedy at the time, but we had people we knew with us, and you just got up every morning and went at it. We managed to carry on, but it was a blow."

The long summer evenings in the Cariboo provided additional daylight hours for more than just maintaining the equipment. The next order of business was building several portable cabins to house the Ainsworth family and their crew. After all, they may have had to truck in everything from water to food and fuel, but there was no shortage of three-quarter-inch rough lumber at their disposal.

"These weren't just plain old shacks. They were really good 10- by 20-foot cabins with good windows and doors," recalls Allen Ainsworth, who was 11 years old at the time. "After a while we even put some wallpaper in them. We had some good sawdust-burning stoves, and the cabins were warm in the winter. They were built on skids, so if you were going just a short ways you hooked onto the cabin with Stan Halcro's D4 Cat and just towed it a couple of miles. If the distance was longer, you'd get the cabin onto the back of the lumber truck and move it that way."

The long work days and the isolation of the Meadow Lake campsite made it difficult to keep in touch with the outside world, not to mention staying abreast of changes in the

Stan Halcro with his D4 Caterpillar was responsible for transporting the logs to the various milling sites.

lumber market. Unbeknownst to David, commodity prices had been dropping to the point where someone from Clinton Sawmills was forced to pay a visit to the camp just as the summer was drawing to a close.

"So we were out there and all of a sudden, out of the clear blue sky, they came out and said to us, 'We're very sorry but the market's gone so bad that although we won't shut you off, we can't pay you $20 a thousand anymore. We can only pay you $12 a thousand.'

"Now that was an awful slap in the face," says David, "and for a day or two we were still carrying on, but everybody was looking pretty blue."

At those reduced prices, the chances of staying afloat were slim to none, especially without the resources to tide an operator over until the market picked up again. With their prospects looking grim, David suggested to Albert Friend and Stan Halcro that they take a trip in to 100 Mile House that weekend to look up a fellow named Laci Kardos, who had dropped in on their operation a few weeks back.

Kardos had immigrated to Canada from Europe a year earlier and was a friend of Leslie Kerr, the owner of Lignum Ltd., which operated a planer mill based in Williams Lake.

"This fellow Laci had come out to see our mill at Meadow Lake one day. I remember he just came wandering right out of the bush," says David. "He left his car wherever it was as soon as he heard the mill and just headed toward the noise instead of driving around the rest of the way.

"I had stopped the mill to sharpen the saw and was down on my knees ready to give it a few good licks when he came over to talk to me. I was sharpening away there and he was

The Ainsworth crew drove north through 100 Mile House, right past Herb Auld's gas station.

asking me all these questions — how much are you getting paid? and so on.

"He said he had some logs near 100 Mile House in a timber sale, along with a bunch of logs in piles out in a field that had to be sawn up. He said our mill would be excellent for the job; we could take it right out there and saw them where they were. I told him we were getting paid $20 a thousand and he said he'd pay that. Then he went away and I didn't see any more of him."

Herb Auld in 1992.

Travelling north up Highway 97 to 100 Mile that weekend, the Meadow Lake trio located Laci Kardos and spent the Saturday "running all over the woods, looking at his logs and his timber sale."

Kardos's offer of $20 a thousand still stood, and although the men had their doubts about how long the job would last, they decided to load up their equipment and move operations to the 100 Mile area, about 40 miles up the highway.

The first things to go were the cabins. Although they were built to fit on the back of a flatbed truck, they were over-height as far as public roads were concerned. To avoid bringing down half the telephone lines in the South Cariboo, someone had to follow along in a pickup, armed with a long pole to hold up the wires as the cabins passed underneath.

With the cabins safely relocated to a site back in the woods north of town, Albert Friend and Stan Halcro went on ahead, moving their own equipment. Next in the procession was David driving his McCormick Deering tractor, towing the portable sawmill, with the board edger tied on behind that. And following him was Susan driving the pickup, accompanied by the two boys.

The owner of the Imperial gas station, a long-time resident by the name of Herb Auld, stood out by his pumps watching in amazement as David rolled through 100 Mile on a quiet autumn evening. "Herb Auld was standing there by the side of the road in front of his service station, watching me pull this little train on my way to 103 Mile, and I waved at him," recalls David. "He probably knew it was a mill I was towing, but he had never seen one like that before."

Spring breakup at the 103 Mile site in the early 1950s.

By that evening, Susan and the boys were already at the logging site near 103 Mile. Darkness was falling and Allen Ainsworth wanted to make sure his father knew where they were.

"Susan and I had bought the boys new bicycles as a kind of inducement to make them happy to come," says David. "I turned off the highway and was on the trail out into the woods, and there was Allen riding out on his bike with his light on. He had come out all that way and he was going to show me the way in. I'll always remember that. He wasn't a very old kid when he did that."

That day when the humble little wagon train journeyed up the highway from Meadow Lake was a defining moment in Ainsworth history. 100 Mile House, which was just a dot on the map at that time, would become the centre of activity for Ainsworth Lumber over the next few decades, not to mention the home of David and Susan, their three children (including Catherine, who would be born in 1958), and the next generation of Ainsworths as well.

As David has often said, referring to that day, September 1, 1952, "It's how it all began."

Compared with the west coast forest industry, the industry in B.C.'s Interior region was in a state of relative infancy when the Ainsworths arrived in the Cariboo. Some 2,000 bush mills were in operation throughout the province's Interior, supplying several dozen planer mills. Around 100 Mile House alone, an estimated 250 portable bush mills dotted the landscape, each of them sawing Douglas fir logs and sending rough lumber of varying quality to a handful of planer mills for custom finishing or dressing. The Cariboo's fir trees were particularly dry, which kept the shipping weights down even when the lumber was shipped green.

In other cases, the rough lumber was trucked to the nearest PGE railway siding, where it was shipped to Vancouver for finishing or sold abroad in its rough form.

Small bush mills had been available for several decades from countless manufacturers around North America. The more portable models with steel frames mounted on wheels, such as the Jackson Lumber Harvester, didn't appear until the 1950s. The bush mills were generally operated by small crews that could pick up and move on short notice, depending on the availability of timber to be cut.

The gold rush mentality of creaming off the most desirable stands of Douglas fir and leaving the rest behind certainly spawned its share of fly-by-night operators. The most unscrupulous of the bunch would simply neglect to pay their log suppliers or other creditors, deliver a final load to the planer, then quickly set up operations in another district, sometimes under a new name.

Up until the early 1950s, the provincial government's method of allotting the timber supply through small timber sales and auctions did a reasonable job of sustaining this multitude of bush mill operators in the Interior. Despite their numbers there was enough timber to go around, although the wastage from inefficient sawing and the approved practice of leaving behind smaller-diameter tree tops was horrendous. In some instances, more than half the amount of timber harvested was wasted through sawdust, slabs and poor-quality sawing techniques.

As timber supplies tightened, competition increased for stands of large-diameter Douglas fir trees — the species of choice. Bush mill operators now faced an uncertain future, and the procedures around timber allotment grew increasingly erratic. The situation came to a head within a few years, and the outcome played a pivotal role in the future of David Ainsworth's fledgling operation.

The more immediate concern, however, was getting things up and running at the 103 site.

Albert Friend with his portable mill, and Stan Halcro with his D4 Cat, got to work in the bush on some property close to a parcel owned by local rancher Bill Day, where Laci Kardos had a timber sale. David, meanwhile, took his portable mill to a field near Forest Grove Road where locals Gus Anderson and Maurice Teslo had several piles of logs ready to saw into rough lumber, also arranged by Kardos.

It turned out to be a poor choice of location and a frustrating start to this new phase of activity. The field was on a hillside, which made it difficult to set up the portable mill, pile the lumber and load it for delivery.

"It would have been much better if he had left the logs in the woods," says David. "But we went out there and somehow managed to cut them."

Fortunately, that job was completed in a couple of weeks, then it was over to the 103 site where he could get back to cutting standing timber brought in by Stan Halcro.

Susan Ainsworth had returned to their home at Cultus Lake to put the boys back in school, but their enrollment there was short-lived. An employee's wife had been hired to come up and take Susan's place as camp cook, but the workload proved to be too much. Within a few days Susan, Allen and Brian were on a Greyhound bus, bound for the Cariboo once again.

> *The night that Susan came back up on the bus with Allen and Brian, she got into town at about two in the morning. I guess I was late or something in picking her up, so she had some time to spend over at the 100 Mile Cafe. There was this waitress there — kind of a gossipy type, I think — and she starts saying things like, "Oh my God, you're not going to put your boys into the 100 Mile school!" She started telling them all these terrible stories about what happens to kids who go there. Well, she had Brian so worried he didn't want to go to school the next morning. Allen just wanted to go and straighten things out. So Allen went to school and came back, and Brian asked him how it was. Allen said, "Great," so then Brian went the next day.*
>
> — DAVID AINSWORTH

Laci Kardos, along with partner Bob Malkin, called his company Cariboo Manufacturing and had set up a small planer mill on Exeter Road at 100 Mile, which provided much-needed access to the PGE railway. He was buying all the rough lumber that David could produce, which in those days averaged about 15,000 board feet a day.

At this stage, the vast majority of David's time out in the bush was spent running the portable sawmill. It was heavy, tiring, monotonous work involving long days of riding a tandem carriage back and forth through the logs.

The board edger and gang saw, with the family's cabin and cookhouse in the background.

Steve Richards taking care of laundry chores.

The Ainsworth crew takes a break.

There was plenty of muscle involved, but the job also required a steady watchfulness to ensure a consistent quality of lumber was being produced, with a minimum of waste.

David's days started early in the morning before the crew was even out of the bunkhouse, first filing the saws and then looking after any equipment repairs that were required. More often than not he'd be out there sawing late into the evening when conditions permitted.

"You couldn't really leave the portable mill," says David. "If I did have to leave it for somebody else to take over the sawing, they might have been fairly capable, but they just didn't give a damn."

Determination and self-reliance were essential attributes if a bush mill operator planned on staying in business longer than a season or two. Without the resourcefulness to look after all the equipment repairs, the resulting downtime meant a loss in production. When that happened, crews quickly disappeared in search of a steadier paycheque.

David's mechanical abilities saved the day on more than one occasion.

Work at the 103 site came to a standstill one winter day when the crew arrived to discover that the engine on the board edger had somehow burned up overnight. It appeared that a canvas tarp used to cover the engine had been set on fire by a hot exhaust pipe. Normally, the tarp wouldn't have been laid over the edger until the pipe had cooled, but someone had been in a hurry.

"We had about four guys working for us at the time, and as soon as they saw that burned-up engine, they jumped in their cars and headed back down to Chilliwack," says David. "They probably went away saying to themselves, 'Well that's the last of this; this will be the stone that fixed the oxen.'"

With little recourse but to get to work, David tore out the fire-damaged engine and hauled it back to the cabin they used as a dining room — the only place warm enough to do the repairs.

"I had to put the engine right up on this big heavy table we had in there, made out of some planks. I don't think Susan was too pleased, but you couldn't work outside in that kind of weather with your bare hands."

Once he took the engine apart and determined which parts needed replacing, David jumped in his pickup and headed straight for Chilliwack — the closest location for such items. The parts bill came to a couple hundred dollars, which just about broke the bank, but within four days he had the engine back together, freshly painted, and the edger was up and running again.

"So I called these guys down in Chilliwack and said, 'Come on back to work.' I think they were pretty surprised. Jobs weren't easy to get and we paid regularly, so these guys came back up and we went to work again."

Producing the rough lumber for Laci Kardos was David's responsibility, but the delivery

> *Cliff and George Garrow came out to that field of Gus Anderson's one time to pick up a load of logs. Cliff was driving the truck. The ground was soft and the front end of the load was slightly off-kilter and low, and we couldn't get the truck under it.*
>
> *So we shut down the mill and we're all out there trying to help. We put some 1x6s in front of the deck, and we were trying to hold these boards in such a way so the truck could get underneath the load. George Garrow was up on the deck. It takes a bit in these conditions; your wheels are spinning on the wet grass or whatever, and Cliff had to take a few runs at it while we were trying to hold the boards in place. Finally Cliff got the deck underneath the load, but in all the excitement and the pushing and the pulling, George got his foot underneath a board and it pinched his foot pretty bad. Well, he was hollering to beat hell, and some of us were yelling at Cliff to hold it, stop it. They could both be kind of short-tempered, so Cliff turns off the motor, puts on the brake, gets out, slams the door and comes around there and says, "What the hell are you doing back here?" By this time the rest of us had shut up a bit, and George is screaming at the top of his voice, "You're sitting on my goddamn foot!" He said this in no uncertain terms. He was hurt. So Cliff immediately jumped back in the truck and drove ahead, and they got his foot out. He was hurt pretty bad, though, and by this time he was doing a little bit of foul language.*
>
> —David Ainsworth

The truck was loaded by backing under the rough lumber, which was hand-piled on special gates that gave way as the truck backed up. David was somewhat embarrassed by the "Lumber Harvester" emblazoned on the truck's fenders. "It didn't have to be quite so glaring," he says, "but I guess everyone knew who we were."

was contracted out. It was initially handled by two brothers, Cliff and George Garrow, who had an Austin flatbed truck. The Garrows hauled loads for several other bush mills, as well. When the roads turned soft during spring breakup, David couldn't help noticing that the timing of the deliveries often seemed to favour the other operators, leaving his loads sitting at the landing until better road conditions prevailed.

"There was nothing more agitating than having three or four loads of lumber waiting to get moved," he says. "You couldn't afford to have lumber just sitting there."

By now David had become acquainted with service station owner Herb Auld, who also served as a dealer for various farm implements and vehicles. It just so happened that Auld had access to a shipment of Austin trucks and chassis that had arrived at the Vancouver docks the previous year, but had yet to be sold.

"These were new trucks that had been put out in a field somewhere," says David. "I guess they had sat out there too long, so they were going for a lower price. And so we bought a truck."

Before he could haul a load with his new Austin, he outfitted the bare chassis with a wood deck and the set of rollers that made it possible to roll in under the lifts of lumber.

Now equipped with his own means of delivery, David could reduce his dependence on the timing of others. The Austin's purchase, however, had unforeseen implications.

"It was an improvement, because now as soon as the roads froze a little bit, we could get right out there with a load. We ended up with far less lumber in our yard at breakup, so we were kind of independent then," he says. "But in a way, having the truck made things worse. Not only did I have to run the mill during the day, but now I had to start taking loads in at night, and we were often hauling lumber right into the weekends."

Of course, the more equipment involved, the more potential there is for breakdowns and other interruptions. This new responsibility for hauling the lumber into town brought its own set of challenges. Getting stuck axle-deep on a muddy trail with a full load of lumber wasn't uncommon, nor was wrestling with a reluctant engine in freezing temperatures.

On one particularly frigid night of −25°F, David had delivered his last load of lumber and was driving back to their camp near 103 Mile. Well off the highway now and back in the bush, the Austin decided to stall.

"I got out quickly with a few tools and whipped off the flexible line that goes into the fuel pump," says David. "When I got it open I blew back on it — in those days you didn't worry too much about getting a mouthful of gas once in a while. That seemed to free it up and I put it back together as quickly as I could. I went about a mile or so and damn if that thing didn't do it again. Out I'd go again, take that gas line off and give it a blow. By this time, with those low temperatures and that cold gas running over your hands, you were in pretty bad shape by the time you got it going.

"I ended up doing this about four times that night; it would go a mile, then stop. I was getting so cold because there wasn't enough time for me to get warmed up. The last time it happened my hands were pretty near petrified, and I said if this damn thing doesn't keep going, I'm going to walk the rest of the way. Somehow or other I got it going one more time and managed to get into camp, but I was really fed up with that thing by this point."

With money tight and payments on the new truck to look after, David had no immediate plans for any more equipment purchases. Most bush mill operators weren't exactly prime candidates when it came to arranging financing for equipment purchases or bidding on timber sales. Within a very short time, however, he was one of the only bush mill operators in the region to own a forklift — a significant advantage when it came to loading the truck and moving logs around the site.

The purchase wasn't even his idea.

It was a mucky spring day and Bill Cretin from Kelowna Machine Works had to park his car and walk the remainder of the way in to the 103 site. David was doing some repairs on the portable sawmill when Cretin offered to sell him a two-wheel-drive forklift for just over $11,000.

"I said to him right away, 'It sounds wonderful, but we haven't got any money. I couldn't even give you a down payment right now,'" says David. "I actually tried to discourage him."

Not one to give up easily, Bill Cretin returned to Kelowna and convinced the company's owner, Harold Armeneau, to sell a forklift to Dave Ainsworth on credit, for $1 down. It was a bold move, but Cretin was convinced there was a substantial untapped market for these forklifts in the region. If they could showcase the forklift's usefulness for one operator, others would want one too. Sales to this point had been largely restricted to fruit growers in the Okanagan Valley.

"In later years Bill told me he recommended me to Harold because he thought I would know how to run it, do a good job with it and have success with it — and that I wouldn't beat the hell out of it or bad-mouth it," says David.

This $2 bill was used as the down payment for Ainsworth's first forklift.

David receives the $2 bill from Bill Cretin 30 years later.

Bill Cretin in 1952.

David Ainsworth was the first bush mill operator in the region to use a forklift.

Before Ainsworth workers had a forklift, they moved logs onto the saw's carriage by hand, using a peavey.

The deal was finalized when Bill Cretin returned to the bush with the offer of a sale for a down payment of $1.

"The ironic thing was that we couldn't find a dollar. Here we were out in the weeds and I had a two or he had a two, but we didn't have a dollar. So I actually put $2 down on it."

The story of the forklift purchase would seem to end there, but the conclusion didn't come for another 30 years.

"Bill and his wife were on holidays up at a guest ranch in the 1980s, and he phoned me to say they were coming through town and that he wanted to have lunch. I said that would be great, I hadn't seen him in a long time. So Susan and I went down to the Red Coach Inn and we had a nice lunch, then out comes this $2 bill. Bill had folded up this $2 bill nice and neat after we made the forklift deal and tucked it into the corner of his wallet. His wife would say, 'For god's sake, why do you keep that $2 bill in there?' And he would tell her this $2 bill had quite a story attached to it.

"So he gave me the bill and said, 'I'm giving this to you now — my wife has been bugging me about this long enough!'"

The forklift itself was called a Timber-Toter, which consisted of a used Ford truck with the body rebuilt so the driver's seat faced backwards to the open end. It had a new engine and a roof, but no doors or windows, and the loads were lifted with a one-piece boom. "It was a great help but it was only two-wheel drive, so if you didn't have a load you were pretty helpless because there was so much weight on the front axle," says David. "You could get stuck on a wet leaf. Not long after they came out with four-wheel drives, and then we changed and got another one."

In addition to loading the truck, the forklift also proved invaluable in moving logs around the landing, especially the tree-length logs that Stan Halcro was now bringing in with his Cat.

"Having a forklift was probably one of the single biggest improvements to the portable mill operation," says David, "especially when the four-wheel-drive models came around."

The Timber-Toter wasn't the only piece of equipment uncommon for a bush mill site. David also purchased a Marathon gang saw, which was set up on skids and powered by a Ford V8 engine. The gang saw was about two feet wide and had ten vertical saw blades that cut rough lumber far more accurately than any portable mill of the day. Three-sided cants from four to ten inches thick with the square side up against a fence would be run through the saw, producing precision-cut 2×4s, 2×6s, 2×8s and even 2×10s. The end result was better use of the timber resource: more boards per log, less sawdust — and the precision cuts meant less planing was required.

Work at the 103 site lasted a year or more, with David now overseeing a crew of six. By this time, Albert Friend had packed up his sawmill and moved back to the area around 70 Mile.

There were a few small jobs around 111 Mile to look after, but the next major move for the Ainsworth operation was to a timber sale in the Edwards Lake area, about 25 miles northeast of 100 Mile House. Once again, all the equipment was loaded and moved out, followed by the cabins. There was little problem with the over-height loads during this move, since the cabins were being taken to a more remote area with fewer low-hanging phone lines to worry about.

During the next three or so years at Edwards Lake, the pace of change quickened considerably, although the core activities remained constant: sawing, piling and delivering lumber; keeping a crew busy, fed and paid on time; and keeping the equipment running.

"With the gang saw going now, we had gotten to a point where we were producing three, four or five loads of lumber in a day, which was fine because with the forklift you could set them aside," says David. "The first load was always put on the truck and the rest were set aside. Our object was to get it all delivered to the planer in town. Every night at quitting time I'd sharpen the saws, we'd have a quick dinner, and then I'd leave in that truck with a load for the planer mill. Then I'd come back and load myself again with the forklift and take another load down. We got to where we never had less than three full loads to go, and then every second day we'd have four loads — maybe five if some one-inch boards developed.

"There was always something to do in the day, so you'd be hauling these loads in at night. As soon as I got one load hauled and the kids were in bed, Susan would sometimes come along, to keep me awake as much as anything. I could haul three loads a night pretty easy from that place and be done by midnight. Then I'd be up at 5:30 the next morning and back at it again."

Living at the work site in cabins as they did — David and Susan slept on the bottom bunk, boys on the top — the line between business and family was thin if non-existent.

"Time never dragged on, that's for sure," says Susan Ainsworth. "I actually liked being out

> *One of the most frustrating things with the portable mill and the gang saw was when you broke down and needed a $5 part, you could be waiting for hours. The only phone was in 100 Mile House at this little telephone office. It was run by Mrs. Watson — her son Carl later came to work for us — and she was just too efficient sometimes. You'd go to make a call to one of these places, and it would be evening because when else are you going to make it? And you'd be sitting on the bench and waiting in line. Then she'd let you make your call. If you had two calls, you'd have to finish one, then go back to the bench and move your way up the line again. Mrs. Watson was so efficient, though, she'd plug into your line and say, "Are you through?" If you took time to draw a breath and say, "No, I'm not through. I'm waiting to see if the fellow on the other end has my part in stock," she'd think the line was dead and then she'd cut you off so the next person could use the phone. Too bad for you. Then it was back to the bench and the end of the line. I spent a lot of time in that office.*
>
> — DAVID AINSWORTH

in the woods. It was so nice at night, with the moon out. We were born and raised on the farm so we were used to being out. I didn't feel as if I was stuck out there because I didn't come from the city. And I always knew that no matter what happened up in the Cariboo, we still had our house at Cultus Lake.

"The boys just loved it out there, but I'm very glad I didn't have our daughter Catherine in those days — I don't think I would have taken a girl out into the bush like that, although I probably would have had no choice."

At the busiest times and before a cook was hired, Susan was preparing meals for up to 12 men each day, some of whom boarded in the cabins on site, and others who had homes in the area. "The men were happy if they were fed well," says Susan. "We had a friend in Chilliwack who had a butcher shop, and once a week he would send us this big box of frozen meat on the bus and we'd get it the next morning. I'd start with the fish first, then the liver next and then the steaks, and so on, down to the ground beef by the end of the week."

Sons Allen and Brian were now attending elementary school in 100 Mile House, but even their daily ride into town was linked to the family business.

At roughly 13 years of age, Allen was as tall as Susan and had already learned to drive the Austin truck. Setting out for town in the morning, Allen would drive the truck with a load of lumber out to the highway, a distance of about 10 miles. Then Susan would take the wheel for the remainder of the trip to Laci Kardos's planing mill on Exeter Road.

"Once we got to the yard, Allen would dump the load and then I'd drop the boys off at school," says Susan. "Then I'd go pick up my water and fuel. After that I'd get my groceries and get back home to make a hot lunch for the men. At three I'd go back into town to pick up the boys and then get home to cook supper."

Growing up in the midst of a portable bush mill operation, there was never a time when Allen and Brian weren't around logging and sawmilling equipment. After school and on weekends, they'd do everything from piling and tallying lumber to driving the forklift — even when their feet barely touched the pedals.

Road trips to the coast were often delayed by construction throughout the Fraser Canyon.

Stan Halcro pulling the cookhouse to a new 103 Mile location.

Cabin contents had to be well secured to survive a trip behind the bulldozer.

> When we were moving the cabins and the cookhouse from site to site, I would usually pack things really well before a move, and then I'd get out of the cabin and Stan Halcro would pull it with his Cat. Then I'd go in and put things back together and get it set up for cooking. This one time we were only going to be moving a short distance, less than half a mile, so Stan said, "You don't need to pack anything. I'll be careful." Stan and his Cat being careful! Well, we had bowls on the counter — fortunately they were aluminum — and boy, they started taking off. Mildred [Stan's daughter] and I were supposed to be watching them. We were holding these things down but we could no more hold all these things down than fly. We had kettles and dishes going all over the place and we were screaming and hollering, and of course Stan couldn't hear us. We sure laughed about it after.
>
> — SUSAN AINSWORTH

David, shown at the front of the International 170 truck, taking a break before another move.

The cabins' final resting place was on Exeter Road, where the family lived until 1956. Allen is on the forklift and Susan and Brian are beside the vehicle David called the "tin jeep." Once while he was driving the jeep on Highway 97 at 111 Mile, a deer bounded into the side of the vehicle, which caused "far more noise than actual damage."

> I bought a small light plant for the cabins, so we had lights, but they damn near caused us a problem. Allen and Brian had a little room of their own and, as kids would do, they rigged up a long cord with a light bulb so they could read in bed. When Allen got up one morning while the power was off, he must have plugged it in and then forgot what he was doing or threw the light bulb on the bed when he was looking for something. We came back that afternoon and sent him out to start the light plant. That light bulb lying on his bed started a fire. Nobody had noticed that the light was on the bed; the boys were out giving me a hand with the forklift. Fortunately, Susan was there in the other room, but it was a pretty good fire going by the time we found it. It really wised us up, and we had a little lecture then.
>
> — DAVID AINSWORTH

As Allen recalls, "We had no radio or television out there so we had to find something to do. In the winter, Brian and I would go find some little frozen pond, often no bigger than a room. We'd just skate around and around, jumping over logs and patches of grass sticking up through the ice."

In warmer months, chopping down poplar trees would often keep the boys amused.

"Allen and I would do this for entertainment because we had an axe," says Brian. "These trees would be four to eight inches in diameter, and there would be a little patch of them, maybe 20 to 30 trees. So we would undercut them and back-cut them and then cut down one tree on the side, and it would knock the whole patch down. For us it was a big thing; it was huge. I guess that's how we learned to domino fall. Of course, it scared the hell out of my mother when a patch of trees fell down unannounced."

Jack Lefferson, who had been with the Ainsworths since their Fraser Valley days, grew to be like an older brother to the boys, taking them hunting and fishing when time permitted — and attempting to keep an eye on them.

"I think Susan tried to keep them out of the bunkhouse," says Jack. "She thought they might get to smoking cigarettes or listening to the wrong kind of stories."

Long weekends or a few days off for the Ainsworth family usually meant an all-day drive back to their home at Cultus Lake for everything from laundry to chores to banking. The Fraser Canyon section of the Trans-Canada Highway was under constant construction in those days, so lineups and delays were common, with the trip often taking 12 hours.

"It was hot in the summer and miserable in the winter," says David. "Those trips could be a real chore."

These were lean times for the family — and would be for some years to come — primarily because every spare dollar the company earned went back into equipment. Continually looking for ways to reduce waste and improve productivity was already becoming a hallmark of the Ainsworth operation.

> *I took over the lease for the 100 Mile Chevron station in 1955. One day I came home and said to my wife, "My god, there's a woman who comes into the station, and she is always dressed so nicely and wears a dress." That was uncommon for the Cariboo. I said I couldn't believe it. That was my first impression of Susan.*
>
> *We had a shop and did some work on their vehicles for them, and right away their good business practices stood out. Accounts at businesses were tough to carry in those days. With all those 250 or so little mills around, we were pretty cautious with accounts. There were a lot of accounts that didn't get paid, but the Ainsworths' account was always paid on time, every bit of it. That was very important to them, and that was Susan's part.*
>
> — CHUCK SHAW-MACLAREN

As Jack Lefferson puts it, "David could do almost anything the business needed. Even in the early days, he wanted his equipment kept in good shape, taken care of. There were probably 100 other sawmills right around 100 Mile, and if the operators sold a bit of lumber, a lot of them would probably stop at the bar and maybe not come back to camp for a day or two. If Dave had a few extra dollars, it went right back into the outfit to improve things or get new equipment. And they damn near went broke a dozen times doing it."

Although it was standard procedure for bush mills to sell their rough lumber to a planer mill, there was always potential for disagreement over the amount they'd get paid. Differences often arose over the weight of loads or the percentage deducted because of culls.

Rather than haggle with Laci Kardos's Cariboo Manufacturing over such issues, David decided that planing his own lumber would be a smarter way to go. Within a year of arriving at Edwards Lake, he had purchased a Newman 4-knife planer.

"You could run the planer with three men," he says. "I'd usually be feeding it, and then we'd have an extra guy out there at the back doing the trimming and the grading."

Allen Ainsworth was in his mid-teens by this time and recalls that first planer vividly.

"Dad set the planer up on skids and ran it off a GM two-stroke diesel. The planer mill worked okay for what he needed to do, but he did have a problem with these laced belts that ran the top head, the bottom head and the side head. He made these light-duty belts into endless belts by using some alligator clips and running a pin through them. The trouble was, though, that these clips would tear out on an hourly basis. There was a twist in the belt because it had to go from a big horizontal pulley to a small vertical pulley. It was murder on the belt, but he had to use them like that for some prolonged period of time until he got the proper endless belts.

"Somehow or other he got all this working, even with these light-duty belts at first. He had a little trim table, which he built himself pretty much out of nothing. There was no conveyor either. Every 15 minutes you had to take all the trim ends and heave them over into a box."

In addition to acquiring his own planer, the other significant development at Edwards Lake was the replacement of the Jackson Lumber Harvester with a newly designed rig called a Bach Mobile Sawmill. The Bach Mobile was designed by Art Bach and was sold by the Robert E. Malkin Lumber Company of Vancouver.

The sawmill required six men to operate it and included a head rig, carriage, edger, power plant, slab conveyor and sawdust blower in a single unit. With an average output of 20,000 board feet a day, it easily surpassed the performance of the Jackson Lumber Harvester.

Stan Halcro, meanwhile, continued to bring in the logs and also purchased his own Bach Mobile around this time.

With only a small crew at their first Exeter Road site, David could often be found running the Newman planer.

The Marathon gang saw.

The Bach Mobile portable sawmill required a crew of six to operate, but it could produce up to 20,000 board feet of lumber a day.

Near the end of the Edwards Lake stint, the 1953 Austin was traded in for a six-by-six, tandem-axle, front-end-drive truck. "It was a bit of an old beast," says David, "a rebuilt thing from the army, but we needed something with more power. We were going through one of the worst breakups I've ever seen. There was no road out there — just a trail, and in the midst of this breakup the bottom of that trail just fell right out. If you had to pull onto the grass or somewhere to get around a bad spot, you needed that extra power."

It was early summer in 1955 when a distraught Laci Kardos showed up at the Edwards Lake site first thing in the morning. He had been in the process of building a new planer mill on Exeter Road in 100 Mile when disaster struck.

"They'd been working on this thing for months and it was just about ready to go when they had a fire," says David. "He came out to our place begging me to take our planer into his site because his new planer had burnt down. Somebody was welding or something and they lost the whole thing. He knew the mill was going to be out of commission for many weeks, but he had all this lumber piled up and ready to go, and a crew was there with cabins all around."

For David, the decision to move his planer onto an Exeter Road site wasn't a difficult one. In fact, he figured it would be only a matter of months before he'd have to do it anyway, given the economic benefits of being located next to the railway.

"So the plan was to move the Newman planer in and plane some of Kardos's wood and hopefully some of ours. We started hauling rough lumber in there and putting it in a yard

David files the saw while Jack Lefferson (top) leans on the cants to assist their passage through the gang saw. Lefferson's job paid $1.75 an hour at the time.

Another load of rough lumber is delivered to the planer mill at Cariboo Manufacturing. Loads were rolled off the back of the truck by the sudden stop-start motion while in reverse.

down the road a ways. We put the planer up in Cariboo Manufacturing's yard, which was just off the road in a clear spot."

With the focus now on clearing out the backlog of rough lumber piled around the Kardos site, David was running his planer virtually non-stop, producing everything from 2×3s right up to 2×10s. What he hadn't counted on was the immediate hue and cry that went up over the noise generated by the planer's power plant. "We ran that Newman with a 671 GM motor without a muffler on it, and it was a noisy devil," he concedes. "Out in the woods at Edwards Lake it wasn't too bad because it ran during the daytime. The problem here was that we were running two or three shifts and there were cabins up by the road, which was the only clear spot in the yard for us to set up. The cabins were for the men and their families who were working on Cariboo Manufacturing's planer. Well, some of the women got really upset with me because we didn't shut that damn planer down. It was the only way to get ahead.

"I said to them in no uncertain terms, 'I'm sorry but your fellows are all working here on this planer, so at least they're getting paid. I'd love to move this outfit the hell out of here and shut it down quiet, but this planer is why your men are working.'

"There wasn't much of an argument, and they left us alone after that. If we were going to be there any length of time we probably would have put a good muffler on it and kind of toned it down a bit, but it was definitely one of those 'damned if you do, damned if you don't' situations."

Moving the planer into town also had implications for the Ainsworth family. Rather than drive back and forth from Edwards Lake to town every day, David trucked several of the portable cabins into 100 Mile, just up the road from where the planer was stationed, and set them up in an adjoining fashion to afford more space. The family's days of living at a remote site out in the bush were over.

The new location was certainly welcomed by Susan.

"Moving into town and being able to set up better living quarters was very nice," she says. "I got relieved of doing the cooking for the crew and driving the boys into school each day. I didn't have that kind of work to do each day so now I could zero in on billing out the cars, helping to do the tallies and things like that."

After several months of planing rough lumber for Cariboo Manufacturing, it became clear that another change was in order. For starters, Laci Kardos's planer mill had yet to be rebuilt. All of David's energy was going into planing someone else's lumber, and there was no end in sight. In the meantime, his own rough lumber, still being cut by a crew out at Edwards Lake, was steadily piling up. There weren't enough hours in the day to get all the work done. The little Newman planer could barely handle 15,000 board feet in a shift.

"It was quite a time," says David. "I finally said we've got to get out of here. We were getting a little pay from Cariboo Manufacturing but only for the planing. We had too much of our own lumber piling up down the road and we weren't getting any of that dressed."

So the planer was moved a bit farther up Exeter Road to the site where the family's cabins were located, near what is now Sollows Road, so David could concentrate on dressing his own lumber. His problems, however, weren't entirely solved. Production on the portable mill out at Edwards Lake had dropped off noticeably now that he was no longer there to run the saw and oversee the crew himself.

"It just went downhill," says David. "The people I had left to run the Bach Mobile kind of slacked off when I wasn't there, and pretty soon they weren't producing as much as they were before. They were taking lots of coffee breaks and they never really got the thing going. We gave them a contract price as a bit of an incentive, but it didn't solve the problem."

Dissatisfied with the production of cants coming in from Edwards Lake, David struck a deal with Stan Halcro, who was supplying the logs and running his own Bach Mobile there at the same time. Halcro would take responsibility for running both of the portable mills, although the Ainsworths would maintain ownership of theirs.

"So Stan still had the logging for both the mills and he managed to get the two crews side by side, and they worked together," says David. "It wasn't an awfully good arrangement but he did a pretty good job of it. Stan was really murder on machines — especially his pickups, which he just beat to hell — but he could get people to work on time, and the results were pretty good."

With the issue of production at Edwards Lake now resolved, David could turn his attention back to his planing activities in town. It was 1955 and sons Brian and Allen, barely into their teens, were already beginning to play a role in the family business. Both boys could handle a forklift and now spent many of their after-school hours and Saturdays moving cants and loads of lumber around the yard.

"By the time we moved into the site near Sollows Road they were both skilled and capable forklift drivers," says David, "probably better than any people we could hire. If you had an extra machine you could always put one of them on it and they would come and work the holidays and evenings. They were pressed into service a lot, mostly on the forklift, and of course kids just love to do something like that. Out at Edwards Lake they were too young for the portable mill — that was pretty tough physical work. But they were there every time they were needed and never seemed to complain too much. Everybody's kids were expected to work back then, especially if you lived on a farm or ranch."

The Ainsworth boys were paid the regular rate of pay, which Susan kept track of, David adds. "When they went away to school, they had a little fund set up to cover costs in Williams Lake, Kamloops or Vancouver."

> *I never had any problems working with Dave — he always treated his crew real good. He was always very patient and a good man to learn from, quite meticulous. There were never any tantrums or swearing and jumping up and down. If you did something wrong you were told about it and that was the end of it. He was always good that way. And he and Susan worked well together; I would say it was more of a partnership.*
>
> *It always felt like a family outfit. In the spring of 1955 the breakup was a month-and-a-half long and we still weren't working. Generally it was two to three weeks. For some reason my girlfriend was going south and so I decided to leave too. When I told Dave I was going, it was like leaving home.*
>
> — JACK LEFFERSON

The money, however, wasn't always used as intended. On one notable occasion, Allen returned to 100 Mile from Vancouver, showing up on a brand new motorcycle.

"He had spent all his money on this new motorbike, so when he got home he had to ask Susan for more money," says David.

As the log cants came in on a steady basis and the production of dressed lumber gradually increased to a couple of railcars a day, David knew one ingredient was still missing: direct access to the PGE railway line. At this stage, the dressed lumber was still being tallied at the planer, then loaded on a truck and hauled to a public loading dock a couple miles up the road, where it was hand-piled into the waiting railway cars. Three of Ainsworth's earliest employees, Red Inman, Ron Semley and Guenther Dast, looked after the loading duties on a contract basis.

"I was keeping my eye open all this time for a chance to get up on the rail side, and eventually that came to pass," David remembers.

Farther up Exeter Road, about two miles west of the railway station, two local operators, Carl Pelkey and Joe Danczak, had a parcel of land where they sorted their logs. Although the site was farther out Exeter Road than any other operation, it did feature access to a railway siding.

Pelkey and Danczak did their logging in territory west of the site and dragged the wood in using logging trucks fitted with arches. The peelers for veneer were shipped to Vancouver, while the sawlogs were sent for processing at the former Western Plywood mill on Exeter Road.

"Carl Pelkey eventually left the partnership, so this site up there on Exeter Road became available when he and Joe Danczak split up in 1956," says David. "The site was leased from Bridge Creek Estate. It had a bit of siding and when they quit they left it to us. The only reason to be there was to be on the rail — that was the goal of moving to the Exeter site."

During their five or so years of travelling around the South Cariboo from timber sale to logging site with equipment and cabins and family in tow, David and Susan Ainsworth were never really sure where the next step would lead them. Now their lumber operation had found a permanent home, and it was literally at the end of the road. There would be growth and expansion, the inevitable downturns in the market and innumerable challenges still to come, but through it all the Exeter site would serve as the heart of the company's activities for the next three decades.

The Ainsworths had arrived.

Alan Lefferson (Jack Lefferson's brother) with Allen and Brian Ainsworth in 1954.

David's father, George, inspects the planer mill during a trip to B.C. in the early 1950s.

Chapter 4

LODGEPOLE PINE PIONEERS

Allen Ainsworth (above) and his brother Brian were experienced forklift drivers at an early age.

Despite the immediate benefits of having a railcar siding closer at hand, it took the better part of several months to move the company's operations up to the Exeter site. For one thing, keeping the money coming in was paramount, so the planing activity and lumber deliveries had to continue throughout the transition. In the meantime, David had enlisted the services of local millwright Walter Paterson to construct new log decks for the planer, along with other modifications. But there was still plenty of welding and fabricating left for David to do.

"Moving onto that site was definitely a step-by-step affair," says David. "I still had to run the planer at the other site, so I was running back and forth, in the evenings and whenever I could find time, to help Walter Paterson set things up."

The North American economy was in a bit of a slide, and the lumber markets were following suit. As usual, money was tight, but David hadn't lost his determination to upgrade equipment whenever possible to improve recovery and increase production. He managed to arrange a deal on a Newman 500 6-knife planer, still a small size by industry standards, but a large improvement over the 4-knife model he was currently using.

"Getting the Newman 500 was a big step for us," says David. "It was a 'fits-all' sort of machine. Later on, when we really pushed it, we could get considerably more out of it than it was designed for — up to a maximum of 200,000 board feet a day. That's with a good infeed, good outfeed and lots of people. It was a big difference."

As useful as the Newman 500 was, transporting the planer to 100 Mile House was no small feat.

"Allen and Brian and I went to Ashcroft with a pickup and one of those tandem trailers to pick it up. It was so hot in Ashcroft that day that it would just make you faint, and this planer had to be pulled out of a railcar. We took along jacks and chains and come-alongs and all sorts of things. Then we had to inch it out of the car, literally had to pull it out with come-alongs and get it onto the trailer. Man, it was a tough job. I was so glad to get into that pickup and get a little air moving and drive home."

One of David's first innovations in product quality at the Exeter site was trimming the ends of the lumber before it went into his new planer, using trim saws set up at the planer's infeed table. The difference was strictly in appearance, but it meant that Ainsworth's finished lumber had ends that were devoid of any roughness or stubble. Paying such attention to edge trimming, along with a properly done eased or rounded edge on the board, took extra time and effort. But it was improvements like these that would soon come to differentiate the Ainsworth product in an increasingly competitive marketplace.

David also purchased a Marathon II gang saw, or sash gang, specifically designed to saw cants rather than small logs. The Marathon II was slightly wider but shorter in height than its predecessor, and its shorter saws made it superior for resawing cants. It was run off a Cummins diesel engine.

I started in the bush when I was 15. The country was surrounded by small mills, and some of them were bad small mills — dangerous places where guys didn't get paid or fed properly. Nothing but moose meat in the winter — you'd get a lot of that. And people who came and worked a couple of days and left; we called them bunkhouse inspectors.

It was mid-afternoon July 23, 1957, and it was very hot weather. I had heard a rumour that this Ainsworth outfit was hiring, so I thought I'd have a look and took a drive up there. I just walked onto the site and there was a forklift there and a small office, along with a planer and a gang saw. They were just moving to the Exeter site. There were a couple of legs and boots sticking out from under this forklift, and I walked over and asked another fellow there — a logger friend of Dave's I think — where Mr. Ainsworth was, and he said he was "over there, under the forklift." *This guy comes out from under the forklift wearing coveralls with a cigarette sticking out of his mouth. I said, "I heard you're hiring and do you need anybody?" He said, "Yeah, I need a guy right over there," pointing to the planer. I said, "When do you want me to go to work?" and he said, "Right now." So I just got my gloves out of the truck and went over and went to work at the planer. Your training session only lasted about 30 seconds. It was pretty much, "There it is, get at it." If you had the common sense you just went ahead. When I finished that day, Dave was standing on the office steps and he said, "Well, what do you think, are you going to come back tomorrow?" I said, "Yeah, I'll see you in the morning." At that time, there were about nine employees.*

— CARL WATSON

Such capital expenditures may have seemed risky at the time, but David's penchant for upgrading machinery was always coupled with Susan's close attention to the company's finances.

"In the early days with that new equipment, everything we got was on some kind of conditional sale," says David. "Susan made sure that the cheques went out every month and that we never got behind in our payments. That way we were always okay for another deal. On some of these things we didn't have to put much money down, but we managed to keep the payments up even if we went short of something for ourselves. And that happened more than a few times."

With his new planer in full production, David immediately took advantage of the rail siding by designing an additional belt off the end of the planer's chain that made it possible to run the finished lumber straight into the boxcar. By now Guenther Dast and Ron Semley had developed an efficient system to hand-pile the lumber, probably based more on self-preservation than anything else.

"Guenther and Ron got to be really good car loaders," says David. "They were organized, and they had to be. They took turns with the boards coming out, and they came out fast. These guys could handle it pretty well, though, and it was too bad if they didn't because those boards just kept on coming."

With 10 or so employees on the payroll by this time, David was still filling in wherever an extra man was needed, taking a hands-on approach to everything from the forklift driving to equipment repairs to construction around the site.

The example he set didn't go unnoticed. Carl Watson was only a teenager when he started with Ainsworth back in 1957, but he was no stranger to bush camps and upstart sawmill operations.

"The secret to the company's success was the fact that everybody pitched in," says Watson. "If someone fell down, there was somebody there to pick him up. If the forklift was broken down and the wrenches were at the shop, someone was booting it up the hill to get them. If a battery was dead, someone had his hood open on his pickup truck and they were over there with the jumper cables. This is the sort of thing that was instilled. I was just a kid, 19, and I really learned this stuff."

The region's small sawmill and planer operations were also notorious for sending employees home when equipment broke down or production was unexpectedly halted. David took a different approach. He knew it would be easier to keep good employees if the work was steady, even if it meant paying a crew for cleaning up the yard.

"There was always work. David didn't send people home," Watson adds. "The whole time I worked there, even after I got into the management part of it, the thing was to keep the crew working and keep them happy. Most people were relying on a steady paycheque. They appreciated the work, even if it was cleanup."

For some jobs, such as moving a railcar along the siding, the team effort was essential. Normally a certain type of winch would be used but David didn't have one, so five or six men would have to inch a loaded railcar down the track with a special ratchet-driven jack.

> When Allen and Brian were still in school, they bought a '53 Chev car, brought it home, and then decided that they had to lower the suspension. This car also had a standard shift on the column, so they took this shifter and switched it over so you shifted gears with your left hand instead of your right. This made the whole shifting pattern backwards.
>
> One day Susan was in the office on a Saturday and she had to drive this '53 Chev because her car was in the shop or something. I was there looking after the planer, doing some greasing and so on. So she's driving the kids' car and makes it down to the rail station and then stops at the stop sign. When she goes to take off she stalls the car right in the middle of the railway tracks because of this backwards shifting pattern. She can't find first gear on this car and she is furious.
>
> I think the gearshift got changed back after that.
>
> — CARL WATSON

The Marathon II gang saw was designed to saw square-sided cants rather than small logs.

As production gradually increased and another employee or two was added to the payroll every few weeks, Susan Ainsworth's role in the company also expanded accordingly. After all, they were now official: the Ainsworth Lumber Co. Ltd. was incorporated in 1957. A small, sparsely furnished cabin at the Exeter site served as the company's office and as a storage place for spare parts. From here, Susan looked after the payroll and day-to-day finances, as well as tallying loads of lumber and preparing the bills of lading for shipment.

"I had a time book and a synoptic and a ledger and I kept everything down in that. I was self-taught, but common sense told me what to do," says Susan. "When I did the payroll in the earlier days, I had to go down to the post office and buy the unemployment insurance stamps. All the men had a little book and each week, when they got paid, they got a stamp in their book that showed they had worked. The system actually worked okay."

> *We used to joke that the further you went up Exeter Road, the tougher the people got, and we were the last ones up the road.*
>
> — DAVID AINSWORTH

As for tallying a load of random-length lumber — it wasn't for the mathematically challenged. The mixture of pieces could include everything from 2×4s up to 2×10s, with lengths ranging from 8 feet right up to 24 feet. With the boxcars of the day holding some 30,000 board feet of lumber, David improvised formulas to aid in the process of tallying a car.

"You had to develop these skills as you went along and I don't think I could do it very well," says David, "especially if Susan was away and I might have to do some typing. It was quite a little process to fill out a bill of lading that detailed the number of pieces and the board feet in the car, and then of course the price and so on. Then you had to prepare separate bills of lading for the PGE and the American railroads if you happened to be shipping to Seattle or Milwaukee or somewhere. You had to learn a lot of things in a heck of hurry without too many people helping you."

Between production and construction, the Exeter site was the scene of almost non-stop activity during these formative years. One of the first additions to the site was a permanent burner for the wood waste that came from the planer and gang saw.

Most of the building activity, such as putting up a roofed enclosure around the planer, had to go on without impeding the lumber production. As David was famous for saying, "You've got to put lumber in the car. There's no money unless the lumber is in the car, and nothing else matters until that car is delivered to your customer."

Although a shop and other small buildings were springing up as needed — locals Al Blackstock and Stan Findlay played key roles in the construction — the new site was still a no-frills affair, stuck out at the end of Exeter Road with no power or even a phone. The stretch of dirt road from the Exeter railroad station to the site was so bad that employees had to be shuttled back and forth in a trailer when heavy rains turned the road to gumbo.

"When we first moved up there it was an awfully wet spring with terrible rain every day," says David. "The road was just freshly made and it didn't have any gravel on it. It was so greasy that we had to keep people off it, so we rigged up a trailer behind Chuck Shaw-MacLaren's old army jeep. The men would park at the station, and that jeep would go back and forth from the station to the mill because the road was so damn bad."

It would be several more years before the site had hydroelectric power to run the equipment and lights, so David pieced together a generating centre using the old planer's GM 671 diesel engine, along with a diesel engine stripped out of a D8 Cat. The setup gradually grew to include two generating sets that provided the necessary power, but not without a sizeable explosion if the operator failed to adhere to a careful startup procedure.

"Using two generating sets was not uncommon, but you had to coordinate them precisely when you threw the switch on," says David. "The cycles from both generators had to be in concert by slowly bringing up the rpm. If they weren't synchronized properly, the one that was running would force the other one into synchronization — with a helluva bang. It scared the hell out of a few people who went in not knowing what they were doing. If the wrong guy got in there, these generators could flash fire and kick the switch back on you. There was a fair amount of danger connected with these things."

Hydro power was slowly making its way out Exeter Road — the Canim Lake Sawmills site closer to town now had electricity — so David figured his time would come, even if it did take until 1960. In the meantime, all of his equipment — the planer, trim saws and gang saw — were installed with electric motors in anticipation of the day he'd have power.

> *We had to get the first stud mill finished so we could get it started the next day. It was getting kind of chaotic. There were lots of logs in the log yard, all of which had to be paid for, and we needed to get the mill going. It was getting to be night and I still had to weld these four-inch channels into place on a new incline conveyor that would carry waste right out to the burner. I'm down on my knees on this conveyor and when one welding rod was gone I'd just stick in another one. Then it started to rain. Pretty soon it started to thunder and lightning and everything else. It was a helluva storm. I thought, "This is stupid, I should quit." But no, I'm not going to quit when I only have another 50 feet to go. It was so stupid to stay there. It was thundering and lightning and I'm welding on a steel conveyor in the rain. This arc welder is giving me a shock every time I use it. It would go through my knees and my legs, which wasn't too awfully bad because it was low voltage, but it still gave you a jolt and made you jump. The water was pouring off my nose and I was just like a stubborn child, saying, "The hell with it, this isn't going to make me stop." It was a good demonstration of how sometimes you could really get quite desperate.*
>
> — DAVID AINSWORTH

"The hydro power came at some expense," says David. "We were kind of half ready for it, and we had to foot the bill. They didn't want to put the power in if the mill was going to go out of business. They didn't have much faith in us at that time. Eventually we got some of our initial investment back, but we had to foot the bill upfront."

With operations shifting to the Exeter site and the future looking somewhat more certain — the power commission's assessment notwithstanding — David and Susan decided it was time to move the family out of their small cabins and into something more permanent. Over the course of a year they finished construction on a modest duplex in 100 Mile House, and in 1956 the Ainsworths were one of the first families to take up residence in the village.

At this time, 100 Mile House was little more than a hayfield fronted by the Cariboo Highway and a few businesses strung out along the route, along with a two-room school. The land was actually owned by a British nobleman, Lord Martin Cecil, who left England in 1930 to look after his father's 15,000-acre cattle ranch, called Bridge Creek Estate.

The area was still sparsely populated but its colourful history reached back a century or more. Miners and prospectors with dreams of striking it rich first passed through the Cariboo in the late 1860s, en route to the gold fields of Barkerville, about 250 miles to the north. Stopping houses were established along the trail, and with the town of Lillooet as Mile 0, 100 Mile House was one of many such outposts along the way. Despite its location on the main thoroughfare heading north, however, 100 Mile was far less developed than surrounding communities such as Forest Grove and Lone Butte.

Charter members of the 100 Mile House Lions Club, organized in 1955. Back row, from left: Don Dewitt, Ernest Whalley, Gordon Trusler, Stan Halcro, Art Eversfield, David Ainsworth, Roger Quirin, Melvin Monteith. Front row, from left: James Bruce, Chuck Shaw-MacLaren, Arnold Cooper, unidentified Lions International representative, Harry Nelson.

It would be several years before there was a full enclosure built for the gang saw at the Exeter site.

Now, as the forest industry entered a new phase of development and more permanent jobs were created, there followed increasing demand for housing and all the various services. Even though the land was still privately owned by Lord Martin, 100 Mile House was primed for growth. Early residents such as David and Susan, as well as the town's businesses, actually leased their lots until a vote on the village's incorporation passed with a slim majority in 1965.

There was no high school in town when David and Susan moved in, so sons Allen and Brian would soon be attending classes in Williams Lake or the Okanagan much of the year. During weekends and summers, however, they could still be found working at the Exeter site driving forklifts, tallying lumber, loading boxcars, helping with construction, sawing 250-pound cants into railway ties bound for the U.K., and performing a host of other duties.

The boys also had a new sister with the birth of Catherine Ainsworth on August 22, 1958.

To all appearances, the Ainsworths were holding their own in an era of rapid expansion in the Cariboo forest industry. Portable bush mills were still a dime a dozen sawing cants and rough lumber in the area, but very few of them had made the transition to a planer and gang saw operation with access to a rail siding.

As the Ainsworths' production increased, however, their supply of timber for producing random-length lumber was growing tighter with each passing month. This same era of expansion in the Cariboo forest industry had led to an overall shortage of quality Douglas fir logs over 12 inches in diameter. The spruce was also affected, though to a lesser extent.

Historic 100 Mile stopping house burned down in 1937.

Passengers travelled up and down the Cariboo Highway on this "BX" stagecoach, operated by Barnard Express.

Freight wagons pulled by horses were still a fixture at the 100 Mile House store in the 1930s.

100 Mile Lodge was constructed by British nobleman Lord Martin Cecil in 1930.

100 Mile in the 1940s consisted of little more than a store, a post office and some ranch buildings. Highway 97 wasn't paved until 1952.

The townsite was still a hayfield in 1949. Long-time mayor Ross Marks worked for landowner Bridge Creek Estate.

Foundation goes in for the community hall in early 1950s. The Ainsworths' duplex was built about 100 yards away, soon after this photo was taken.

100 Mile's first village council. Back row, from left: Chuck Shaw-MacLaren, Russ Fraser, Cas Kopec, David Ainsworth. Seated: Ross Marks.

100 Mile House in the early 1960s. The Ainsworth mill is located five miles out Exeter Road, curving left at top.

When 100 Mile House was incorporated in 1965, Ainsworth had approximately 70 employees.

The Ainsworth yard around 1959, when approximately 45 employees were on the payroll.

In earlier years, anyone who applied to the Forest Service to log a stand of timber would usually be the only bidder. When the Cariboo's choice stands of fir and spruce began to disappear, competition for the remaining timber heated up.

Timber auctions were now becoming chaotic affairs, where even the most established operators had to compete vigorously if they wanted to keep a steady supply of logs coming into their mills.

Bidding was sometimes driven to unrealistic levels with the sole intention of saddling a small operator with a sale or stumpage payment he couldn't afford. Outside interests with no intention of logging could pay a deposit and participate in the bidding. Then they would threaten to drive the bids up unless a cash deal was offered by an operator who really needed the timber. In other cases, two bidders for the same timber sale would call a recess, leave the room and cut a private deal. Upon their return, one operator would get the sale at a low rate, then the two of them would work the parcel together.

When the amount of timber logged in one year reached or exceeded the area's annual cutting limits — as was the case in the 100 Mile area — the competition for the fir and spruce grew even fiercer.

"The fact is that the competition got awfully tough for the fir and spruce," says David. "It was very, very ruthless. There were several instances where bidders drove the stumpage up to ridiculous heights. Then the Forest Service would disallow the sale because they knew nobody could afford to cut the timber at those prices. Six months later the sale would come up again, and by God these bidders would drive it up again — they had made up their minds that they were going to prevent another operator from getting this timber. And the large operators did that. We ran into it."

As if conditions for procuring timber weren't bad enough, David's situation grew even more desperate following an ill-conceived directive from the Ministry of Forests regarding the volume of trees each operator was allowed to cut. The Minister of Forests at the time, Ray Williston, felt the area was being over-logged and wanted to conduct an inventory review to determine the volume of merchantable timber. As part of his review, he ordered all the industry players — which included small operators like David Ainsworth — to stick to their current levels of harvesting for one year and not exceed that cut.

The following year, with his new inventory figures in hand, Williston announced that all Cariboo timber firms should have their harvesting quotas reduced by at least 54 per cent.

This ended up hurting the Ainsworth operation more than anyone else.

What the minister didn't know at the time was that virtually none of the other industry players had obeyed his directive the previous year to not exceed their annual logging quota. In fact, all the other firms had promptly gone out and doubled their annual logging activity.

When their quotas were subsequently reduced by 54 per cent, they were still at the same level of cut they enjoyed before Williston's order. David Ainsworth was the only operator in the area who followed Williston's "hold your cut" instructions. When the minister's 54 per cent reduction went into effect, David actually lost 54 per cent of his original quota.

By this stage in the late 1950s, Stan Halcro had traded up for a larger Bach Mobile and continued to supply cants for David's planing operation. He and David still partnered on various timber sales, but they were too small to amass any quota that would ensure a level of long-term supply. As the months wore on, David had to rely increasingly on fir and spruce purchased from private landowners.

"I saw an end to fir trees being available for me," says David. "That's when we got the idea that we better try to find something else."

That "something else" was lodgepole pine.

Although the forests around the South Cariboo and further afield were home to substantial volumes of lodgepole pine, the trees had no commercial value. In fact, the pine was actually considered a weed species. Lodgepole pine often grew in pure stands, especially in the vast areas west of the Fraser River, but boasted an average diameter of less than 12 inches. For the logging and sawmilling equipment of the day, anything less than one foot in diameter was considered too small to process efficiently. Much of the pine grew in a crooked fashion, as suggested by its botanical name, *Pinus contorta*.

The species' biggest drawback, however, was its moisture content. Lumber made from Douglas fir could be shipped green, while spruce could be air-dried in the yard before being loaded onto boxcars for delivery. Air-drying gave the product stability and prevented discolouration and decay.

Lodgepole pine was another story altogether, and in many cases no amount of air-drying would yield satisfactory results. David's gamble on turning this weed species into merchantable lumber was a long shot, but the company's options were limited.

"We knew it was small wood and that it needed to be dried, but with so much competition for the fir and spruce we decided that we had to do something with the lodgepole pine if there was going to be a place for us."

First off, there was a long list of questions to be answered: which product to make, where to sell it, how to market it, and whether such a product would be commercially viable over the long term. These were just a few of the issues.

David and Susan visited a fellow by the name of Chester Cotter, who was a sales manager with the Vancouver lumber wholesaling firm of Battle and Houghland Ltd. Chester Cotter acted as a sales manager for the Ainsworths, as well as for a number of other Cariboo lumber firms.

Sketching out an initial plan of attack, they first decided on a product. Because lodgepole pine was a relatively small, sometimes crooked tree that required significant drying time, they felt that 2×4 studs, 8 feet in length, would be a prime candidate for production. If the timber quality was poor, it was easier to get two 8-foot boards out of a log rather than the longer lengths of 12, 14 or even 16 feet.

One of David's initial ideas was to produce a 1×6 pine board to be used for interior panelling. Although a few panels were produced, there was a similar product already on the market. Sales of the Ainsworth panel were slim to none, and production died a quick death.

Chester Cotter in 1992.

"The panels on the market at the time had a plywood backing. It was a very good product and we would have had to go up against that," says David. "Rather than spend the extra money, we just made studs. The pine panels never came to pass."

By the late 1950s there were already several species of studs being manufactured and sold in the United States: Douglas fir, hemlock, cedar, white fir or balsam, and ponderosa pine. When it came to demand and price, studs made from lodgepole pine ranked dead last. Their only producer, in fact, was a company called Inland Empire, which had operations in Montana, Colorado, Wyoming and Idaho.

The next step in their plan was market research, so Chester Cotter embarked on a fact-finding mission to gauge the customer response to lodgepole pine studs.

Chester died in 1996, but an article he penned in the late 1970s neatly sums up his experience: "The trip to investigate sales possibilities for lodgepole pine studs was probably my most memorable sales trip. And since 1958 I have made many sales trips to various parts of the world. Some of my contacts had no previous experience with lodgepole pine but indicated a willingness to try the product if it was good, without problems and was competitive with other studs. On the other hand, some of the wholesalers were very vocal about their dislike for lodgepole pine studs. I recall some comments which were not exactly

With an enclosure around the gang saw, employees were partly shielded from the elements.

encouraging to me. Such as, 'Don't talk to me about lodgepole pine studs,' or 'If you represent a mill making lodgepole pine studs, get out of my office,' or 'It's the worst stud I've ever seen.' I finally found someone in Detroit who would comment on the bad reputation which had been gained by lodgepole pine."

The story told to Chester dated back to the Second World War. There was a great demand for lumber, so Inland Empire ramped up its production of studs for armed forces barracks and home construction to meet the demand. Unfortunately, they had an unorthodox — and ultimately faulty — method of drying the pine cants and rough lumber. The end result was a stud that was shipped dry on the outside but wet on the inside.

Once installed in walls and partitions, the drying process would continue, much to the horror of everyone involved. Tensions between the dry and green sections would cause the boards to twist, warp, bow and contort, leading to great bulges in the Gyproc or other sheathing materials.

As Cotter discovered, "There were complaints and claims. Building inspectors forced building contractors to remove the crooked studs, all of which meant a loss of time and money. Small wonder that contractors, retail and wholesale dealers were down on lodgepole pine studs. Armed with this mixed reaction, I returned to Vancouver and reported to my employers and to David Ainsworth. We knew that it would be a tough road to penetrate the U.S. stud market with lodgepole pine. The last thing we wanted was for some distributor to say, 'Get these studs out of here.'"

The decision to go after the U.S. stud market with lodgepole pine certainly carried risks, but the Ainsworths saw little choice if they wanted to remain in business. It was the same story for Pinette and Therrien Mills Ltd., another Battle and Houghland client in Williams Lake, about an hour's drive north.

The conveyor chain running from the planer to the burner periodically came loose and went straight into the burner.

A former bush mill operator from Ontario, Gabe Pinette, along with his relatives Dollard and Roger Therrien, had started up a sawmill in Williams Lake in 1953. They were also running out of fir and spruce and began the transition to pine around the same time as Ainsworth.

Cat operator Joe Danczak recalls the day that David told him and Carl Pelkey that he was going to try using lodgepole pine.

"Pelkey and I were up at our Exeter rail siding and we were shipping big fir logs as peelers. We used to load them into these gondola cars. These cars only went so high and the logs were piled higher, so we'd use lodgepole pines as stakes on the side of the car to keep the peelers from falling out. One day Dave came up and said, 'I'm going to make lumber out of these stakes; that eight-inch stake is a good-sized log for lumber.' We kind of looked at him and thought, well good luck. We didn't say anything. We couldn't see how it would work; it just wouldn't pay with that sawmill carriage having to go back and forth. Well, sure enough, he did it. And he made money at it."

The profits, however, would be a long time coming. It was a poor time to be entering the stud market. Top-grade studs made from fir and spruce were selling for between $50 and $60 a thousand board feet. The random-length lumber that David had been producing previously sold for at least 10 to 15 per cent higher.

With an untested product in a skeptical marketplace, David and Chester Cotter knew from the outset that they'd be getting rock-bottom prices for lodgepole pine studs.

"There were lots of derogatory comments at first, and it took quite a little effort to sell these studs," says David. "On one sales trip down to the States some place, this old guy said, 'I don't want any more of that stuff. I had a load one time — you set it out in the yard and when you come back in the morning it's just standing there looking at you.'

"In some cases we said to people who were prospects and were already buying our other lumber, we'll send you a car and if it's not satisfactory, then we'll take it back. And once they tried it, because it sold cheaper, they'd order another car."

Not surprisingly, at one point the price for Ainsworth's lodgepole pine studs drifted down to a low of $46.50 a thousand board feet, which was barely the break-even point.

Anticipating better days ahead, David and Chester Cotter bought a bottle of vodka.

"We were going to keep that bottle on store for the day we sold a car of lumber for $70 a thousand, which was a way out of line."

Although it wasn't foreseen at the time, the decision to switch to lodgepole pine would have major implications for just about every phase of the Ainsworth operation: acquiring, logging and hauling the timber; scaling the logs; drying the rough lumber; adapting sawmill technology to accommodate small logs; and developing products. In many cases there was little or no precedent to go on. David's knack for innovation and his determination would have to see him through.

One of the first issues he had to deal with was ensuring a steady supply of logs. As David soon learned, just because he had switched to lodgepole pine didn't mean his timber supply problems disappeared overnight. The pine grew in abundance — it was considered a weed species and wasn't even included in the Forest Service's inventory for the region. And yet larger competitors such as Blackwater Timber and Canim Lake Sawmills didn't want it to go to Ainsworth.

It all boiled down to quota. Ray Williston had directed the Forest Service to phase out the

David Ainsworth did much of the grader work during the company's formative years.

chaotic market-based auction system and replace it with a timber quota system. So-called established operators would receive a timber quota based on their amount of logging over the previous three years. Timber auctions would still be held, but the established operators would receive bidding privileges on their quota of timber, virtually guaranteeing their annual supply and weakening the position of competitors.

Having decided to go after lodgepole pine, David started out by bidding on sales of pine scattered around the 100 Mile area.

"These were small sales that we put up because we didn't want to attract too much attention. We had about 20 of these sales all over the district. They were deliberately intended to be in areas where there was nothing but pine, or very little but pine. That way we didn't think we would have too much competition from people who would be resentful if they thought we were taking any of the timber."

David's competitors had other ideas. They figured that if someone was now interested in lodgepole pine, this former weed species would suddenly hold more value and the Forest Service would eventually include pine in its regional quota calculations. Therefore, every sale that Ainsworth obtained meant less quota for them. The result was a strange state of affairs not unlike the previous high-flying auctions for Douglas fir.

David would show up at an auction for small sales of lodgepole pine, only to be outbid by competitors who had no intention of actually using the pine. Their only concern was adding to their quota of timber.

Ironically, David initially had assurances from Forest Service staff that lodgepole pine wouldn't be included in anyone's quota. That would make it a far less desirable commodity and keep the prices at a level he could afford. His competitors' hunches were correct, however, and it wasn't long before the pine was included in a newly established quota system.

Joe Danczak skids logs to the landing on his D4 Cat.

Knocked out of the bidding process once again, David's only recourse was to approach the winning bidders and offer to cut their newly acquired stands of lodegpole pine. And it worked. Although he now had timber at a price he could afford, it was a far better deal for his competitors. They were selling previously unmerchantable trees and building up their quota at the same time.

The situation soon came to a head when the Forest Service's new quota system, which now included lodgepole pine in the inventory, was officially put into practice. Once again, David Ainsworth came out on the short end of the quota system, through no fault of his own.

In a previously published interview, forests minister Ray Williston described the whole scenario, going back to the occasion when David followed orders and held his annual cut in check while others proceeded to double theirs.

> *I screwed Dave Ainsworth up twice. I thought we were over-cutting in the Cariboo so I naively went in and said to industry, "Just hold your cut right now because we're gonna take an inventory of this and see what it is." So in the next year or two, everybody in that 100 Mile area had doubled their cut except Dave Ainsworth, and Dave Ainsworth stayed exactly where he was. When I finally got the inventory straightened out and had to cut everybody 54 per cent, pretty near every one of them came back to where they were when I said stop. But poor ol' Dave, he lost 54 per cent by standing still. I really goosed him.*
>
> *So then Dave came to me and said, "Nobody's cutting this pine around here. I'm gonna take a crack at it so put up a pine sale for me and I'll see if I can do it." I put up a pine sale and as soon as I put it up — we didn't have it in quotas — some of the bigger guys took a look at this and thought, "Oh gosh, we can't let this guy get started on pine. We're gonna have to bid that up." Well, Dave didn't have any money to bid against these other fellas, so he went to them and said, "Look, if you want to take the pine sale, take the pine sale, but I'll make a commitment to you. I'll cut the pine. I'll do all the business*

and you don't have to worry anything about it. If you want the pine sale in your name, well you take the pine sale."

I put the pine sale in the inventory. And he had built up inventory for these other guys and they hadn't cut a stick. I said, "That's a crooked deal," and this is the second time I've goosed this guy on quota. So I went up and called a meeting in Williams Lake and said I was going to put this volume of pine on the inventory. I was going to divide it up and give quota. But before I gave any quota out I was giving a quota to Dave Ainsworth based upon what he had. Before I started dividing everything I was giving him a quota off the top. If they had any complaints, let me hear them now. I never heard a complaint. The whole industry said, "Okay, give it, he earned it, that was fair." In those days in the Interior we could talk that way with the guys and they all said, "Yeah, that's fair, away you go."

The establishment of a quota system set the stage for massive changes in the Interior's forest industry. In essence, logging rights could now be bought and sold, which initiated a wave of industry consolidation that continues to this day.

Hundreds of bush mill operators were now assigned quota and found it advantageous to sell their logging rights to larger operations eager to solidify their position. Within another five years, the company started by David and Susan Ainsworth would be among just 20 to 25 operations still producing lumber in the Cariboo.

There were still numerous outfits in the 100 Mile area around this time, the largest of them being Canim Lake Sawmills. Brothers Rudy and Slim (and later Walter) Jens started out with a small sawmill at nearby Canim Lake in 1942, then moved onto a site on Exeter Road with a planer mill in 1948.

Before long, the green-end studs became a sign of quality in the industry.

The Exeter mill site in the late 1960s.

By 1957 they were out of bush milling and concentrated on the finishing end of production. In 1958 they put in a gang saw and another planer at their Exeter Road site. By 1962 they would add a drying kiln and another planer, and in 1965 started up a plywood division. They were eventually bought out by Weldwood in the mid-1970s.

Brothers Glenn and Jim McMillan had been supplying lumber to Canim Lake Sawmills with their portable mill throughout the1950s. In 1963 they started up their own sawmill near the community of Lone Butte, about 12 miles southeast of 100 Mile. McMillan Contractors was bought out by Ainsworth in 1978.

The planer operated by Laci Kardos evolved from Cariboo Manufacturing to Kardos Forest Products and in 1962 was sold to Komori Lumber, located near 70 Mile.

Further north, in the Williams Lake and Quesnel area, there were a number of operations making a name for themselves, in addition to Pinette and Therrien.

Leslie Kerr's Lignum mills in Williams Lake and Quesnel already employed 100 people. His planer mill in Williams Lake, a major contributor to the local economy, had been operating since 1948 with a network of area bush mills providing the wood.

Out near the community of Horsefly east of Williams Lake, Harold and Odt Jacobson refurbished a burnt-out bush mill in 1954, added a planer and moved the entire operation into Williams Lake as Jacobson Bros. Sawmills in 1964. The Jacobson brothers grew up in Terrace, operating their father's water-powered mill, then took on various logging jobs around B.C. before arriving in the Cariboo.

Another set of brothers, Bill, Pete and Sam Ketcham, came to the Cariboo in the mid-1950s from Seattle, where they had worked at their father's wholesale lumber company. Bill and Sam visited 100 Mile with the intention of buying lumber from Carl Pelkey's operation up on Exeter Road. When that deal fell through — Pelkey ended up selling out to Western Plywood — they ventured north to Quesnel, where they bought a planer mill and then bush mills in the area. In 1957 they set up a cant and planer mill near Williams Lake, calling their company West Fraser Timber. Sam Ketcham looked after the production while his brothers handled the sales from Seattle.

Rounding out the region's forestry scene in Quesnel was Weier's Sawmills, operated by Herb Weier. Weier's was another of the mills whose lumber sales were managed by Chester Cotter of Battle and Houghland, and would soon be making the transition to lodgepole pine.

While the field of B.C. Interior lumber producers was narrowing, competition for U.S. sales was steadily heating up. Selling studs made from lodgepole pine would be an uphill battle. The lumber would have to stand out — beyond the fact that it was entering the marketplace at rock-bottom prices.

David felt the solution was to focus on product quality, and he quickly established a set of guidelines:

- The product would be well manufactured, with a very good planed finish and accurate trimming.

- There would be no compromises on grade, and it would have a low wane (bark or lack of bark) content, which would enhance the appearance of the stud.

- The studs would be put into a pallet and steel-banded with edge protectors to prevent gouging of the outside studs in the pallet.

- Both ends of the studs would be painted with an attractive colour.

Allen Ainsworth (left) directs efforts to re-erect a stack salvaged from a Prince George planer mill. The equipment was used to heat the Exeter site's kilns with wood waste.

The emphasis on superior quality proved to be a successful strategy. Within a decade or so, the Ainsworth stud with the distinctive turquoise end-paint would be considered the industry's gold standard, but it took a great deal of effort and determination to get there.

One of the first hurdles to overcome was ensuring the studs were properly dried.

"I visited a couple of mills in the Okanagan where they were using lodgepole pine for random-length lumber and selling it locally, and they were air-drying it out in the yard," says David. "But there was a constant wind down there and their winters are warmer than ours. And their wood was probably drier than ours to start with."

Clearly, any lodgepole pine studs from the Cariboo would have to be dried in a kiln. But aside from one or two kilns in the Quesnel and Prince George area, these furnaces were virtually unheard of in the region. "Drying kilns in the Interior were like hen's teeth," says David. "There weren't very many of them."

There were Moore-built drying kilns at the coast, but they were generally found in larger mills. For a look at smaller operations, David and a representative from Moore toured several

U.S. states in the Northwest before deciding on a suitable design for drying lodgepole pine lumber. It was a significant expense, even though Battle and Houghland assisted with financing the purchase of Ainsworth's first drying kiln.

"This was a big step for us to take," says David. "A really big step, but it was an essential part of the plan."

Carl Watson was on shift the day that first kiln's foundation was poured.

"We shut down the planer and the gang mill and Al Blackstock came in with a portable cement mixer. They put the forms in for the kiln and we poured the cement. We didn't have any trucks, it was all done by hand, and we didn't quit until the job was done. Then we started up the planer and the gang mill and went back to work."

Although most kilns are fuelled by natural gas, the Ainsworth kiln was initially heated by the exhaust of a three-million-BTU burner that ran on furnace oil. The main reason for going with oil was cost: it was too expensive to bring a natural gas line all the way in to the Exeter site. The oil furnace tended to burn a bit smoky, turning the rough lumber a toasty brown colour. Any discolouration, however, was removed during the planing process.

Despite plenty of expertise provided by Moore when the kiln was set up, the real challenge came in determining the right settings for drying the pine in all types of weather. David would measure and re-measure the lumber's moisture content with a probe, testing the results of varying drying times under different temperatures and venting conditions. The average temperature inside the kiln was 170°F.

The site's new beehive burner was running at full capacity within a short time.

"After a while you'd get some experience with the thing and get to know what time you'd need for different conditions," says David. "You'd be looking for moisture content under 10 per cent, which might have taken 70 hours in there. If you're going to have steady production that you can count on, you've got to have the lumber coming out of there every 70 hours, winter or summer. It was a steep learning curve."

Within two years, production warranted a second kiln, and in 1965 the oil-burning furnace was replaced with a massive hot water boiler that was fuelled by wood shavings from the planer mill. The unit, salvaged from a Prince George planer operation, was 6 feet in diameter and 18 feet long, capable of producing 200 horsepower as a steam boiler before it was converted to hot water. When the kilns were heated with wood waste, the cost was virtually zero — about $1 per thousand board feet.

A drying kiln wasn't the only major expenditure made around 1960. Although the company's Marathon gang saw could readily trim boards to eight feet — the required length for studs — it wasn't the most efficient piece of equipment for the job. David knew that a sawmill specially designed to produce studs would give better log recovery and increase production capacity. So thanks to an $80,000 loan from the Industrial Development Bank (IDB), he managed to purchase a stud mill. The deal marked the first of many between Ainsworth Lumber and the IDB over the coming years. Nevertheless, the IDB initially kept the Ainsworths on a tight leash: as a requirement of this first loan, David and Susan's combined annual salaries were not to exceed $8,000.

> *When we built the green airplane hangar in 100 Mile, most people would have bought some steel, made some trusses and put the thing together. But the old man decided that he could make his trusses out of one-inch and splice them all together. So he laid the pattern out on the ground with a little bit of an arc, and we had to do the nailing. We were up there on Saturdays and Sundays, pounding thousands and thousands of those damn little ring nails to make all these trusses out of one-inch. Those trusses are five layers thick with a 40-foot span, and they're still standing there today. And it didn't cost us much.*
>
> — BRIAN AINSWORTH

The stud mill took several months to build. Consisting primarily of components designed by Marathon, it was built to produce eight-foot boards from small-diameter logs and was constructed over a large V-shaped conveyor that carried away any wood waste to the nearby burner.

The mill was able to trim logs into eight-foot sections and send them down the line in a continuous stream. The pieces then went through an adjustable twin-saw or scragg that turned them into cants. The four- and six-inch cants then went through the double-arbour edger that cut them into 2×4s or 2×6s.

With the edger's fine-cut carbide saws, the new stud mill was able to saw these pieces to an accuracy of one-sixteenth of an inch. Controlling the width of the board to such an extent meant far less wastage when it came time to plane the lumber for finishing, not to mention better recovery.

The final step in this process of log to lumber was directly tied to the quality guidelines David had set out at the beginning: on each pallet, both ends of the studs would be painted with an attractive colour.

Chester Cotter's article provides the details: "While a stud mill was being constructed and the kilns were set up, I investigated colour possibilities with a paint company in Vancouver called Brandram-Henderson. We wanted a long-lasting colour which would not fade, and there was a lot of experimentation with various colours. Finally we decided on a particular shade of green.

"In due course, studs were being made and it was time to paint. I travelled from Vancouver to 100 Mile with the Brandram-Henderson sales representative, Cliff Crispin. There were several pallets of studs set out in the mill yard. Chuck Shaw-MacLaren, who was the shipper at the time, together with Cliff Crispin, painted the ends of the pallets. What excitement, and what a beautiful package. Today, end spraying is commonplace but this was new and different for an Interior mill and the green colour was outstanding."

The green end-paint was more than decorative; it also provided a seal to prevent the board's ends from checking or cracking.

Chuck Shaw-MacLaren recalls his role in painting the very first wave of packages ready for shipping.

"For the first loads, we put the pallet down and got the ends stacked nice and smooth because I needed to roll them with paint on a roller just before we loaded them into the railcars. I had done some spray painting before I worked for Dave, so I soon got a sprayer and started using that.

"At this point we still had to do the spraying outside. If it was raining, we'd put up a blanket on the one end. If the wind was blowing, then anything nearby was green."

As David puts it, "There were a lot of green forklifts — and green forklift drivers."

A decade or two later, when Ainsworth studs were making their mark around North America and in overseas markets such as Japan, the green end-paint grew to be an undisputed symbol of quality within the industry.

Chester Cotter continues: "As time went on and demand for the Ainsworth stud increased, the green colour became very important as a source designation. Buyers knew the names of all the mills in the stud business. Most buyers would not refer to the Ainsworth stud, but would say, 'I want another car of those green-end-painted studs.' I have heard that request hundreds of times. In 1975 the Japanese market began to develop for Interior sawmills as did the U.K. and European markets. At the forefront of demand was the green-end-painted stud from Ainsworth."

As David would soon discover, the switch to using lodgepole pine entailed a complete rethinking of their logging and hauling practices. The region's harvesting system up until the early 1960s was built around the large-diameter Douglas fir and spruce trees. Loggers used chainsaws to fell, limb and buck the trees to length right at the stump. In some cases, axes were used for the limbing when the saws were deemed too heavy.

A steel cable or choker was set around the log, and then tractors or small Cats would skid the pieces to a nearby portable mill. When the skidding distance proved too great, the portable mill was moved closer to the area being logged.

"That's one of the reasons the portable mill was quite successful," says David. "You laid it all out and had landings here, there or wherever to keep the skidding distance down to a minimum, which was the practical way to do it."

The system worked well enough for larger trees, but as diameters grew smaller and the number of pieces multiplied, the costs of logging this way soon became prohibitively expensive — especially when the mills became more stationary and the skidding distances increased.

> My brother Pete and I were out logging at Spout Lake — we had a cabin out at Ten-ee-yah Resort because it was too far to drive. We had a major snowfall and Dave had to come out on the TD24 Cat to clear the road. He had to push out all of Rail Lake Road to get to where we were. There was so much snow that in some places he had to cross-push because he couldn't get rid of it. He got that road opened up and we kept on logging. The job took him all night. Working all night in those days wasn't unusual. You went out and tried to get the job done because there was always something else to do tomorrow. That's how Dave would be. He'd say, "Well let's get it done and then we'll sleep."
>
> — ROBIN NADIN

At the Exeter site, fir and spruce cants from Stan Halcro and other suppliers were still being delivered for processing into random-length lumber, but the bulk of David's attention had shifted to lodgepole pine and eight-foot studs.

The pine initially came in from private sales, but by 1962 Ainsworth was taking on increasing responsibility for the purchasing, logging and hauling of its timber. Fortunately for David, he didn't have to oversee all this additional activity by himself. Allen and Brian were now working full-time for the family business.

Although there was never any pressure from their parents to stick around, both sons had pretty much decided at a very early age that they'd be part of the business.

"Long before I started full-time, I knew I'd be working for the family business," says Allen. "I think mother did express two or three times to us boys that it would be nice to have us stay and work, but that was when we were young teenagers. We never thought of doing anything else. There was so much going on all the time, and it was for yourself — for the family."

By the age of 19, Brian had completed his schooling and had been hauling cants and logs for the family business in a converted lumber truck for nearly two years.

"There was never any real separation between the family and the business in those days; they were the same thing," says Brian. "It was like any farm family living on the farm, only we didn't have a barn — we had a sawmill. Instead of a tractor, we had a forklift, and you went to the mill every day. Working for someone else was something I never even considered."

Allen, meanwhile, was pursuing a forestry degree in 1961–62 at the University of British Columbia in Vancouver.

This industry gathering in the late 1960s included David (back row, second from left) and Brian Ainsworth (back row, fourth from left), Glenn McMillan (back row, far right), and "Fuzzy" Komori (front row, second from left).

Cants from Stan Halcro's operation continued to come into the mill.

"The plan was for me to finish my degree and come home to work for the company. You could become an entrepreneurial businessperson by the time you were 21, so no, there was no hope of me working for someone else. By this time I was unfit to work for anyone else."

Brian concurs: "I don't think we were worse than anyone else, but we weren't really housebroken. We couldn't have worked for anyone else."

In 1962 Allen returned to 100 Mile for a summer job, but a serious car accident derailed his plans for returning to university.

Robin and I had a bucking contract in the early days, and in the summer it was pretty ugly. They skidded the logs out of the bush and put them on the landing, and you'd see nothing but limbs. The trees were all packed in there close together and you could only get at three sides with your chainsaw. You couldn't get underneath. So maybe two-thirds of the tree is getting limbed. It was a common complaint and we were constantly being harassed by the sawmill people over the quality of our logs, and rightly so. But we were doing the best we possibly could. It would be 80 degrees by 10 in the morning, and Brian [Ainsworth] would come out to the bush with instructions from the sawmill that he "better smarten those buggers up out there." Then he'd see what we were faced with and he'd just drive by. He knew that if he stopped, he'd probably get a power saw thrown at him.

— PETE NADIN

He and his soon-to-be wife, Hiroko Uyeyama, were driving north to Williams Lake with friends Larry and Anne Pinkney for a Sunday night drive-in movie. They had just passed David and Brian, who were heading south in the International lumber truck after picking up a grapple in Williams Lake, when an oncoming driver swerved into Allen's truck while making a left-hand turn at the Horsefly cutoff.

Hiroko sustained a broken pelvis and Allen suffered a concussion and a broken right leg that would put him in a cast for the next five months. He decided to forgo school that autumn and remain in 100 Mile, and he and Hiroko were married October 30, 1962.

While the Ainsworth boys were now full-time employees, the business remained wholly owned by David and Susan. To give them an opportunity to build up equity in their own business, Allen and Brian formed a new company called Little Bridge Creek Logging. Their company would serve as the unofficial logging arm of Ainsworth Lumber.

Allen and Brian's first choice of equipment, however, didn't exactly meet with David's approval.

"Brian and I wanted to buy a diesel logging truck and Dad just laid down the law. He said no, that wasn't necessary, we should buy a skidder. So we bought a Tree Farmer for around $10,000," says Allen. "Within months we literally had to have a loader to keep up with what the skidder was doing, so we bought a John Deere front-end loader. Ainsworth then bought its first logging truck, a 1962 Mack diesel, and then Little Bridge Creek Logging bought its first logging truck, another Mack diesel, in 1963."

By 1962 Brian Ainsworth was fully occupied with the hauling activity required to feed the stud mill. He continued to drive a logging truck until at least 1965, when his supervising duties in the bush took precedence. Allen Ainsworth was dividing his time between work in the bush — heavy-equipment maintenance, driving the loader, organizing the day-to-day logging activity — and covering an ever-growing number of bases at the Exeter site.

"There was always something to do," says Allen. "We worked Saturdays. There would be some shaft on a piece of equipment that had to be changed, or something that had to be welded. We were always building or adding on or changing things. Brian and I used the small company shop to build things, like a log grapple for one of our loaders. Another time we built an arch on our '63 Mack truck because we couldn't afford to purchase one from the usual suppliers. We just looked at other people's arch trucks and built our own. There wasn't much grinding involved and very little painting — it was just functional."

When it came to their operations in the bush, the Ainsworths' biggest challenge was developing a cost-efficient method of falling, limbing and skidding the pine trees.

"The transition between fir logging and pine logging was difficult and clumsy," says Brian Ainsworth. "At one point we would drag everything into the landing with limbs on it, then try to limb it all when it was lying on the ground. At first it was just a mess; there were too many pieces to deal with. But you had to keep trying different things because what used to work with fir just didn't work now."

> *After that car accident near the Horsefly turnoff, the cast was on my right leg, but within a month or so I was going around and doing a whole bunch of chores, picking up nuts and bolts and all these things. We had pickups with standard transmissions and I would put my cast and leg up there on the seat beside me. I just learned to drive a standard with one foot. One day I had to go and pick up 50 sheets of this thin hardboard, which was really slick on one side. The only building supply was run by the Minatos up at 99 Mile. They gladly loaded the sheets in the back of the pickup truck, but the panels were too long to close the tailgate. So I drove from there onto the highway. Well, with my herky-jerky one-legged shifting, all the panels spilled out onto the highway behind me. Of course I couldn't pick them up, so the guys from the building supply had to rush out and pick them all up for me. During this period my parents had me looking after Catherine too, who was just about four at the time. That child and I spent all day together every day for quite a while.*
>
> — ALLEN AINSWORTH

A logger's daily output tells the tale: in stands of mature Douglas fir, a two-man team with chainsaws could fell about 100 trees on a good day. With lodgepole pine averaging just 10 inches in diameter, the same team could fell between 700 and 800 trees in good conditions.

Even the direction the pine trees were felled made a difference. With fir it wasn't an issue. The trees were bucked into lengths and it didn't matter which end of the log entered the sawmill once they were delivered. The pine, however, was generally cut to tree length — just the top was cut off — and had to be fed into the stud mill butt-first. Falling the pine in one direction facilitated the choking, skidding, loading and hauling procedures, keeping log handling to a minimum.

This emphasis on felling the trees in one direction led to some unorthodox practices during those early days.

Local brothers Jim and Marvin Higgins, who were among the first loggers hired by the Ainsworths, practised a form of logging called domino falling. Trees that were leaning in the wrong direction were given the usual undercut and back-cut. But by quickly pulling the saw out on the back-cut, the tree was left standing, however precariously. This went on for a number of trees until the brothers found a suitable "trigger tree" at the edge of the patch. With a good breeze blowing in the right direction, they could push the trigger tree over and knock down all the rest in rapid succession, with a minimum of effort.

"Sometimes, if the wind was against you, you could end up with 100 trees all loaded and then this would knock them down at one time," says Marvin Higgins. "Pushing one or two over would get it started. It was like a whole forest going down. Most went in the right direction."

The practice wasn't without its risks: if a straggler didn't topple over completely, there was always a chance it might come down later on top of a chokerman or skidder operator. Not surprisingly, domino falling was against regulations, but officials tended to look the other way. They knew what a challenge this pine was for loggers.

Ainsworth's earliest logging crews also experimented with using pike poles to push the trees over, a practice David had spotted during a trip to Ontario. One man made the cuts with the chainsaw while the other pushed with the pole. The team would switch jobs in an effort to reduce the strain. The technique helped to fell the tree in the right direction and prevented the saw blade from getting pinched, but it was discontinued after several months. Not only was it dangerous, it was tough on the man doing the pushing.

No matter which method was used, logging the pine was back-breaking work, largely because of the sheer number of trees that had to be felled. Forest regulations required that loggers leave a stump no higher than 10 inches. If the tree was only eight inches in diameter, the stump height could not exceed eight inches.

Despite the difficulties, Jim and Marvin Higgins could fell up to 1,000 trees between them in a single day. "It was tough going but we were in pretty good shape," says Marvin. "It was just manual labour, which we were used to. It didn't seem that bad. And we made pretty good money at it."

Logging pine spurred other changes in the bush as well. The usual method of setting choker cables to drag the trees behind a tractor or skidder also had to be revamped.

"Normally, with fir, you'd go out with separate chokers and pick up each tree individually, then bring it back to a concentration point," says Allen Ainsworth. "After that the trees would be loaded together and skidded out. Because with pine there were so many pieces, we developed a method of skidding with slider chokers. If all the trees were felled one way, you could pull out your main line right to the end, set a choker and go off to the side somewhat to set another, setting five or six chokers in all. Then you'd get on your skidder and reel them all in at once. It was a huge improvement over what we used to do with a bull hook and separate chokers."

To cover longer distances when skidding logs, it wasn't uncommon in the late 1950s and early 1960s to use a surplus army truck fitted with a winch and logging arch on the back. These four- or six-wheel-drive vehicles were used to drag bundles of logs, primarily during winter on ice or snow roads where oncoming traffic was a rare occurrence.

Allen and Brian outfitted their '63 Mack diesel with a logging arch and put the long-distance skidding technique to work during a particularly mild winter. A logging site about eight miles west of the mill was well-suited for the trip.

"There wasn't much snow on the road, only frost at night," says Allen. "We'd start at midnight and finish by noon when it began to melt again. When we drove these logs in, they all got sharpened off just like pencils before we were even halfway. There would be wood smeared over the full width and length of the road by the time we were done. It was like a wood road, but it was smooth as glass because you graded it off every time you made a trip."

THE MARK OF AN AINSWORTH EMPLOYEE

One of the more visible signs of the Exeter workforce was a shiny gold hard hat that David Ainsworth introduced in the mid-1960s. Up until this point, employees were responsible for their own safety equipment, even their own gloves. Few if any wore steel-toed boots, and even fewer wore a hard hat. When the occasional employee began to wear a hard hat — one fellow used to turn up with a miner's helmet complete with a headlamp — David decided that every employee should receive one of these aluminum hard hats.

Various workers, proud to have a job at Ainsworth, would wear their hard hats in town almost as a badge of honour. "If the guys wore their hard hats to the bar, of course they stood out like a sore thumb," says Al Smith. "Some guys wore them everywhere. They were quite proud of these hard hats." Proud or not, within a year or two the Workers' Compensation Board ruled that the hard hats didn't meet the current safety standards. The gold ones had to be destroyed and were replaced by the more common plastic models.

Scaling all of the lodgepole pine by hand would have brought the Ainsworth mill to a standstill.

This would be the only time they'd use their logging arch for long-distance skidding, but David was impressed with the results.

"That road we used out west of the mill was just kind of a trail at the time, but if you were hauling downhill in the winter you could take a helluva turn of logs with a truck like that. You could move a lot of wood quite economically because all these pieces didn't have to be loaded onto a truck at the landing. They would be spewed out behind and they'd be 150 feet wide. You'd go around a corner and they'd all go in one way and out the other way. If it was downhill and you could get them rolling, it was kind of fun to drive a truck and haul logs like that. It would get even more exciting if someone was coming along the road with his team of horses and a wagon. You could usually stop in time."

Whether it was felling, skidding or hauling, the sheer number of pieces involved with pine was causing an upheaval in the bush, but that paled in comparison to an even bigger challenge: how to scale so many logs once they arrived at the yard.

Before 1963 the only method of scaling logs in B.C. was hand-scaling. Simply put, a scaler measured the diameter and length of each log delivered on site, then calculated the volume of that log in cubic feet (100 cubic feet equals 1 cunit).

A full load of Douglas fir logs could consist of 15, 10 or as few as 5 logs if it was good timber. The logs were dumped in the yard and each one was hand-scaled to determine the load's total volume. Regulations called for each and every load to be scaled. The provincial government would then use these figures to calculate how much the mill had to pay for logs that came off Crown land over a certain period — a charge called "stumpage."

Hand-scaling 5 or 10 fir logs per load was the norm, simply part of doing business at any mill in the area. A truckload of lodgepole pine, however, contained at least 100 trees and often more. When David saw the amount of time — and log-yard space — it took to scale just one load, he immediately knew he had a major problem on his hands.

"I don't think any of us realized what a chore it was going to be to scale all these logs properly," says David. "It could take two men up to six hours just to do a single load. Not only that, the rules were that you had to measure both ends of the log, even though we were bringing them in virtually tree-length. We had to find some way to deal with this."

There were mills in the U.S. utilizing small timber, so David quickly set out for operations in Montana and Idaho to see how they handled their scaling. He discovered that rather than measuring individual logs, they used ground-level scales to weigh an entire truckload, then convert the weight to volume. The authorities still got their volume figures and the mill didn't have to hand-scale each and every log. It was the perfect answer for lodgepole pine.

"My first question was, why wouldn't that work in B.C.?" says David. "But I soon learned it would be very difficult to get the bureaucrats in the Forest Service to change. They'd say, 'We've been doing it this way for a hundred years and we're going to keep doing it this way.'"

Fortunately for David, he found an ally in Jim Robinson, the Supervisor of Scaling for the district, based in Kamloops. He wrote to Robinson, explained his idea and asked for permission to test weigh scales at the Exeter site. Robinson agreed, and the experiment began.

"Brian and I took the Mack truck to Vancouver and found a place where you could rent these scales, which you mounted on skids. They were designed to go out on the highways when they were doing a highway job. They would put the trucks across these scales and then they'd pay them by weight or by the yard. So we loaded one of these scales on a trailer behind the logging truck. Then when we got back up to Exeter we had a hole dug and a little shack built."

Despite the approval of Jim Robinson for the weigh-scale experiment, David met resistance at every turn from other forestry officials. To make matters worse, he was footing the bill for his scale rental, not to mention all the company time and energy going into a project that would ultimately benefit lumber producers throughout the province.

"These bloody bureaucrats were the biggest problem we had. They were so determined to run everything by the book. Here was something that was so necessary but it was disputed by everybody. They were sticklers for having their rules looked after but very often they really didn't have any idea of what they were doing, and yet they were in charge."

Because weigh scales work on averages, forestry officials insisted that Ainsworth hand-scale 100 loads each of lodgepole pine, fir and spruce to determine the correct conversion factor between a load's weight and its volume.

"They still argued and it made it very difficult for us," says David. "We'd weigh a load and

> *Back around the mid-1960s we had to pour the foundation for the first weigh scales at Exeter, and it all had to be done by hand. We had a small motorized mixer but the gravel and the sand all had to be shovelled in by hand. We had water in drums and we just bucketed the water into the mixer. The bags of cement weighed about 65 pounds and they seemed to get heavier as the day went on.*
>
> *We had some guys wheeling the cement and I was on the mixer all day. I'll never forget it. I was so tired by 7:30 that night that I could hardly get off the ground. I couldn't straighten my fingers out for about a week — they were shaped to the shovel handle.*
>
> *There were a few loads of cement dumped over the side that day, too. We had to walk up these 2x10 planks, and the loads would get slopping around in the wheelbarrow. The more tired you got as the day went on, the harder it was to hang on to those things. So we poured a little extra down around the bottom.*
>
> *We couldn't stop because once you started to pour you had to finish it. That's why we ran so late, but it was all done that day. David was around there steady all day long, and he went downtown to get a case of beer when it was all done.*
>
> *If those are the "good old days," I don't want any part of it!*
>
> — AL SMITH

then that load had to be taken out into the yard where there was room to dump it, and then you'd have to take a machine and spread it all out so the scalers could get a look at every log. It was a curse. People would get tired of walking back and forth on all those logs, but that was the way the forestry bureaucrats thought we should do it. We kept saying there's got to be a better way, but that's how we worked into the better way."

Other mill owners, content with the system as it was, were no help either.

"Everybody said, 'Your idea is a stupid idea.' We'd have meetings at the CLMA (Cariboo Lumber Manufacturers Association) and we'd all stand around in a circle and everyone would be bombarding me with questions and saying how it was not going to work, especially for the bigger timber. Even people like Harold Jacobson — whom I always admired very much — would argue. They had gotten used to the old system, and it was working quite well because a load for them was 5 to 15 logs."

As the months passed and the weigh-scale trials wore on, forestry officials finally started to come around to David's idea. Nevertheless, he was still required to scale 100 loads of each species, given the different densities of the wood.

> When the cants came in from Stan Halcro's bush mill, they came stacked two bundles high on the truck, so you took two packages off together with the forklift. That meant you couldn't see unless you were driving backwards. But when you work in the yard long enough, you get to know where you're going and you think that you don't need to be able to see. One night I was going down the yard with a load and started to get the feeling that there was something on the other side of me, so I lifted the bundles up a bit to peek underneath — and here's the front of Dave's pickup truck and he's backing up like crazy. He said he didn't think I was going to stop. There were a few policy changes after that.
>
> — DICK SELLARS

"The weigh scales were awfully late in coming because this trial period took such a long time," says David. "They insisted on all these loads being scaled. We had to prove it over and over again because they weren't going to take any chances. Fir from one area might turn out to be slightly different from another area. And the weights would be different in winter than in summer when the wood was drier. Finally they had to realize that they had to quit arguing about something that's infinitesimal."

Ironically, the Forest Service turned out to be one of the main beneficiaries of the new weigh-scale system.

"Previously, the scaling had gotten so badly abused by people," says David. "They'd often wait until quitting time to sit down and make the scale. It was always a conservative guess, and the Forest Service knew that they were getting beaten pretty badly. Finally, when they approved it, everybody jumped on the bandwagon, especially the Forest Service. They loved it. They realized this was far more efficient and accurate than anything they had before."

On January 3, 1963, a truckload of logs passing over the rental scales at the Ainsworth mill became the first official load to be weighed in B.C. It would be six months until the province's first permanent scale was registered — at Pinette and Therrien's mill in Williams Lake.

"We had test-scaled our 100 loads of pine and fir, but we were short of spruce and we needed to scale 100 loads of that too," says David. "There was very little of it in the areas we were logging, so it took a long time to get those 100 loads of spruce. By this time Gabe Pinette saw that these scales were going to go, and he actually got his permanent scales installed and ready to go before we did. We were still using the rental scales while we got those 100 loads of spruce. But I say we did all the homework and paid the bill. If we hadn't been able to come up with this system fairly quickly, we would have been dead in the water."

For stud mills using lodgepole pine, and a Forest Service seeking more accurate timber volume measurements, the introduction of the weigh scales proved invaluable. The scales were handled by an employee who was paid by the mill but gained his authority from the Forest Service.

The weigh scales were indispensable when bringing in loads with 100 logs or more.

Even so, the system wasn't entirely foolproof.

David discovered that dishonest logging contractors had quickly found a way to take advantage of this new setup. Some of them would bury rocks or stumps or unusable trees in the middle of a load. Sometimes it was through negligence; other times it was deliberate.

"We had to contend with some of these gyppo loggers who were so dishonest — they would do crazy things like go to the beer parlour and brag about how they could throw a few rocks in the centre of a load and Ainsworth would pay them for anything. There were so many people who thought these scales were a joke.

"We figured it out pretty soon and would tell the scaler to keep an eye on so-and-so. So the next time this guy comes in and there's something wrong, we'd say, 'Okay, wait right here; we're going to put your load down here and have a look.' This would send a message out that if a guy brought in a truck like that again, he was going to have to come back with a whole new load. And we had to do that. We had to send a guy back and say, 'I'm sorry, we won't dump you. You go back to the woods until you load it properly and give us something we can use.'"

In other cases when a substandard load arrived at the site, David would have the logs put aside, then summon the contractor to come in and sort it out. After several hours spent weeding out the junk wood, the contractor invariably paid closer attention to future loads if he wanted to continue supplying the mill.

There was one other bone of contention during the introductory phase of the scales. The Forest Service was still concerned that varying climatic conditions or terrain differences might affect the accuracy of the weight-to-volume conversion. Consequently, they required what mill owners felt was an unjustifiably high number of loads to be hand-scaled for accuracy checks. Gradually, though, as the results proved that such variances were minimal, the ongoing hand-scaling checks were reduced to 5 loads per every 100 weighed on the scales. Even today, regulations require that a certain number of loads per 100 be hand-scaled to ensure the ongoing accuracy of the system.

Ainsworth's first logging truck, this '62 Mack diesel, was like a classroom on wheels for a young Brian Ainsworth, who hauled logs into the Exeter mill until 1965. By the time he was 19, Brian already had several years' experience driving trucks with air brakes. With the arrival of the new Mack, he decided he'd better "get legal" by earning his Class 1 ticket. Trouble was, applicants had to be over 21 years of age. Undaunted, Brian convinced the instructor to let him take the exam. After a successful road test, which included parallel parking the fully loaded rig on 100 Mile's main street, the instructor overlooked the fact that Brian was underage and granted him the licence.

"So there were a lot of things to learn," says David, "but the scales had to come."

An excerpt from a letter written in the mid-1970s by Supervisor of Scaling Jim Robinson sums up the whole experiment: "This weigh-scale method has been found satisfactory for 90 per cent of logs scaled in the province. And without the far-sightedness of Mr. David Ainsworth and his company, this material [lodgepole pine] could have been wasted due to handling costs alone."

Surviving the 1960s was an accomplishment in itself for lumber manufacturers in the Cariboo. For a growing number of operators, the shift to lodgepole pine and smaller-diameter logs was affecting every phase of their operations, from the bush to the mill to forestry regulations themselves.

Staying in business required greater and greater outlays of capital and no small amount of resourcefulness and adaptability. There were no deep pockets at Ainsworth Lumber, so survival was inextricably linked to the efforts and know-how of various individuals, from the owners on down.

Al Smith joined the company in 1962 at the age of 24, after leaving Regal Lumber in Williams Lake. A new afternoon shift had just started up and he was now making $1.75 an hour as an edger — 25 cents more than at his previous job. One of the first things he noticed was his new boss's involvement in every aspect of the operation.

"David was always in there and involved in everything," says Smith. "The big thing was trying to get enough timber and then getting it to the mill. But he would also be involved in the planer mill and the log yard and the logging activity. He'd fill in if somebody didn't show up, and if you needed a hand he would be the first guy to help you do whatever needed to be done.

"We didn't have a lot of spare parts — there wasn't a lot of money around to stock a bunch of parts — so we had to make do with what we had. David always had a feel for the equipment and he never stopped thinking about new ideas, how you could improve on something. That was a big asset he had: how to go about it and find something better. He was always trying something different. We tried a lot of things; some of them didn't work, but a lot of them did."

Employee turnover was high in the 1960s for mills in the region. The industry's notoriously transient workforce was ill-prepared for technological change, and many saw the Cariboo as a brief stopover between larger operations at the coast and Prince George. Numerous employees came and went, but as a second shift was added and the payroll grew, Ainsworth gradually attracted a core group of men whose skills and dedication saw the company through many tough times, as well as contributed to many years of profitability and eventual expansion.

Dick Sellars had already heard positive things about the Ainsworth operation before going up to the Exeter site in 1960 to apply for a job.

Sellars' father, Richard, operated 26 portable sawmills, so he was well-acquainted with the local scene.

"Everybody knew everybody else in the business," says Sellars. "I had heard that Dave had a pretty nice mill compared to a lot of them in the country at the time. When I first met him out at Edwards Lake, he looked to me like someone who was going to go somewhere."

Tired of loading boxcars on the night shift at the nearby Canim Lake Sawmills, Sellars visited the Exeter site one day in hopes of landing a job.

> *I was just new at the mill and Bob Watson was the sawyer on our scragg mill. He had been there about a month already. He had just put new bits into the saw that morning, but he ran the plate onto the feed chain and tore some of them up. It was around lunch time, and because I was the new guy, he tells me to go up and see Susan and see if I can get a new box of bits because we've got to replace these.*
>
> *So I go up there nice and innocent and walk right into the trap. I tell Susan that we need this box of bits, which is worth about $35, and she says, "What did Bob Watson do with those bits he got last night?" I said, "I don't know," and she says, "What the hell are you guys doing down there — eating those things?"*
>
> *So I walked out and thought, "Oh Christ, I'm not going back in there to ask her for anything." I went back to tell Bob and he was already laughing. He had set me up good. Not long after, Dave came down to see where the bits went because Susan had sent him down.*
>
> *Bob Watson is probably still laughing.*
>
> — AL SMITH

> *There was a period there where we started to use what were called hard bits; they were a special hard material and you couldn't file them. It was the going thing in those days for people to steal your bits and take them down to the beer parlour and sell them. And some people did pretty good by this. They were like gold. We had one employee who was walking off with them, so that's the reason they were kept locked up. Those bits were a lot of money and they were just going away as if they had legs of their own.*
>
> — SUSAN AINSWORTH

"I remember David being on his knees in the mill with Gordon Trusler, trying to get the electrics working on the twin saw. He suggested I should come back in a few weeks after they got the mill running. I was looking forward to being able to work with him."

Sellars started on the planer mill's green chain and within half an hour noticed David's emphasis on quality.

"It was pretty evident. The dressing on a board is a little thing, but he wanted it done as good as possible. It wasn't just so-so; it was good dressing. We had a good big planer so that helped, but I'd say we had the best dressing job of anybody that I knew about. It was a case of learning by example. I soon started to realize that this guy was different from the run-of-the-mill people I had worked for before, because he was trying really hard."

In those days at Exeter there was no roof over the green chain and the studs were loaded by hand onto carts outside. The green chain off-loaders had jerry-rigged a little system of rollers and 2×6 ramps that saved them time and effort when rolling the carts out for pickup by the forklift.

"It took a fair bit of practice doing all this, getting everything the way you wanted it, but it worked all right," says Sellars.

Then disaster struck, or so they thought.

"That winter we had two feet of snow one night, so the next day we were all standing around with a hump in our back wondering what the hell we were going to do now, trying to shovel a bit by hand. Pretty soon along comes Dave on the little Cat and he just dropped the blade and pushed everything out of the way. I mean everything — all our little ramps and rollers and parts. And he said, 'Okay, start 'er up!' And I thought what the hell are we going

David and Susan Ainsworth at a forestry show around 1968. David was president of the Cariboo Lumber Manufacturers Association from 1967 to 1969.

Allen Ainsworth (right) with Guenther Dast, working on the fuel bin for the kilns around 1965.

to do now, but there was no fooling around. We had to get to work and keep the cash flow going. I liked that. I thought that's a pretty good way to do things. We weren't just going to wait until the snow melted."

Conversely, if the lumber quality was suffering in a given area, David took the time to coach employees on how he wanted the job done, even if it meant halting production.

"He would just stop you, just stop everything and say, 'This is the way you do it,'" says Al Smith. "He'd stop the whole mill and get us all together. He took the time to do that. You couldn't put out a bad product. You had to stop and fix it."

Although he probably didn't realize it at the time, David's emphasis on quality and improved recovery, as well as his direct involvement in so many of the day-to-day activities, had a lasting effect at the site.

As Al Smith describes it, "After a while we had a real team here. We didn't call it a team like they do today, but it was a team effort just the same. We were very fortunate that we were able to keep a lot of guys for many years, some well over 30 years, or in that range. They started here and worked their way up. A lot of guys were good hands-on people because they were trained by and worked alongside of Dave. He was very demanding in that he wanted it done right. He always stressed that."

Susan Ainsworth, meanwhile, may not have been out front like her husband, but she continued to play a central role in the company's affairs. Whether she was doling out parts and supplies to the employees from the site's small office, looking after the time cards or discussing capital expenditures with David and their sons, Susan developed a reputation for keeping a close eye on the figures and a tight rein on expenses.

Her no-nonsense approach could occasionally test an employee's or supplier's nerves, but no one ever doubted her sense of fairness and honesty. And she wasn't afraid to speak her mind. If someone requested welding equipment or a replacement part, she more than likely knew where it would go and wanted to know what the employee planned on doing with it.

The cutthroat lumber industry was no place for the faint of heart, and Ainsworth Lumber was decidedly an underdog. It took a certain toughness to survive, and Susan's matriarchal presence, both within her own family and in the company itself, had an undeniable influence on the course of events.

"She was hands-on, no question about it," says Dick Sellars. "She looked after the whole front end. If you had a problem you'd go up to the office and she'd deal with it. She had a lot to do with keeping the crew happy, even though things were tough then at the start. One order that we filled was something like $38 a thousand [board feet], loaded in the car. You couldn't really run a sawmill and a planer and a crew on that, but I can honestly say it's the one job I had where they never missed a payday. And I know there was more than once that it was pretty hard, and other people probably didn't get paid so they could pay their crew. That was really respected at the time."

Conrad Pinette, the son of Gabe Pinette in Williams Lake, describes Susan's role this way: "Dave always had the vision and pursued it; Susan had the hammer. So if he convinced Susan, then I guess they would move along. If he couldn't convince Susan, he was in deep trouble. There's no doubt she was a very powerful person in that family organization."

With approximately 130 employees on the payroll by the late 1960s, David and Susan still valued the fact that they knew all their people on a first-name basis. Earlier in the decade the couple would even host the company's annual Christmas party at their duplex in town. When the guest list quickly outgrew their home, subsequent dinners were held at the nearby 100 Mile Lodge and the Red Coach Inn during this period.

Not surprisingly, many employees working for Ainsworth felt a sense of camaraderie and loyalty that set the company apart from other operations, certainly those of a more corporate bent. Speaking at Christmas parties and service awards, David was always fond of telling employees "Your name might not begin with 'A,' but you're all a part of the Ainsworth family." And anyone present knew he meant it.

Seeing that same owner jump in to repair equipment or fill in wherever he was needed set an example for others to go the extra mile when the inevitable challenges arose.

"I worked in different mills in Williams Lake and you were always just a person working there, you didn't mean anything," says Al Smith. "If you left they just hired somebody else. At Ainsworth they were trying to develop a good crew. They felt that if they looked after the people and didn't demand they do something they weren't prepared to do, then they'd have a team. And it worked."

Long-time employee Pete Nadin recalls that one night he was in an accident en route to his home at Canim Lake.

"I got in touch with the office and said I was having problems, so the first thing David does is he phones my wife and says, 'Is there anything I can do? Can I come out and get you? Are you all right?' That sums up the way he approached things. He was the first one in line if help was needed."

While the company was attracting a core group of employees, the Ainsworth family itself was growing during this period. Allen and Hiroko's son, Michael, was born April 16, 1963, followed a year and a half later by Douglas on October 16, 1964. Brian Ainsworth married Hiroko's sister, Eiko, on October 30, 1964, and their son, Kevin, was born June 8, 1965.

It would be at least two decades before they joined the company full-time, but a third generation was now in place.

Each shift at the Exeter sawmill produced enough chips to fill one and a half railcars.

As Ainsworth Lumber struggled to carve out a future for itself in the mid-1960s, sawmill technology was beginning to undergo a massive transformation throughout the Interior.

The changes in the Cariboo were spurred primarily by the provincial government's goal of making better use of B.C.'s forest resources by reducing waste and using more of the wood fibre. Lumber companies at the time had a well-deserved reputation for resisting change — especially when it came to government policies. Trying to develop a greater sense of cooperation among the varied interests was seen as an essential ingredient for improved recovery.

The Social Credit government of the day was keen to expand industrial activity in the province's Interior and north. At the same time, the economy was strong and investors from eastern Canada were looking to B.C. for opportunities.

One of the chief outcomes was the introduction of pulp and paper mills to the Interior, starting with three new operations in Prince George.

As a means of ensuring a steady fibre supply for the pulp mills, lumber producers would now be required to sell wood chips to these pulp operations, which meant that new debarking and chipping technology was needed to produce the chips.

This requirement for chips sealed the fate of the region's remaining bush mills. Their waste was in the form of sawdust and slabs with the bark still on, neither of which was useful for pulp and paper mills.

Leaving this material in the woods to burn or rot was exactly the kind of wastage the government was trying to eliminate. And since only the larger, stationary mills could afford the new debarking and chipping equipment, bush mill operators had little choice but to fold or to sell their timber rights to larger licensees.

At the same time, Ray Williston, Minister of Forests, imposed new logging or utilization standards on the industry, based on Cariboo producers' growing ability to process small logs. It was now permissible to harvest trees with a minimum diameter of 7.1 inches, rather than the previous minimum diameter of 11 inches. At the other end of the tree, tops could now be a minimum of 4 inches rather than the previous 6 inches.

As a result of Williston's new policy, the Interior's merchantable wood supply was increased by 30 per cent simply through the practice of using smaller trees. With more quota now available and the prospect of an increase to their Annual Allowable Cut (AAC), sawmills had a strong incentive to invest in new equipment that could process increasingly smaller logs. This, in turn, led to increased consolidation within the industry as a means of securing the necessary financing for the upgrades.

Ainsworth was approached by several pulp producers to supply chips and eventually settled on an arrangement with Peter Bentley and Canadian Forest Products (now Canfor) in Prince George. The last thing the company needed at this time was to spend $250,000 on a debarker and chipper, but the financing was made available on two fronts.

To ensure a steady supply of chips from the region's sawmills, the pulp mills were offering partial financing or loan guarantees — based on a contract to purchase — to assist in obtaining the new machinery. David and Susan Ainsworth by this time had also established a relationship with Jerry Cotter, who advised the federal government's Industrial Development Bank on various lending proposals. Between the two sources and their own money, Ainsworth was able to swing the deal.

The gang saw at the Exeter site produced edgings, trims and slabs that were previously considered scrap and sent straight to a burner that was overburdened at the best of times. This same waste wood was now the raw material for chips to be sent north by rail.

The pulp mills required carloads of clean chips, which meant Ainsworth's saw logs first had to go through a debarker to remove any stones and dirt picked up in the bush or the log yard. The company acquired a Cambio 26-inch debarker, one of the first machines of its kind. Having clean logs for the first time not only made chip production possible; it also reduced wear and tear on other mill equipment.

"Getting into chips was big for us at the time," says Al Smith. "Before we had the debarker, if you hit dirt and rocks in the log, the bits could come out of the saws and then

The chip pile could grow enormously depending on the market's fluctuations.

you'd have to stop and change them. You could lose quite a bit of time, especially in the spring and fall if you were logging and mud froze on the logs. We'd have to look for rocks and dirt and chisel it off by hand with an axe before putting it through. Otherwise you could damage the saw."

Once the debarked logs were sawn into lumber, the waste wood was sorted by hand and thrown into a belt conveyor that fed the chipper. The chips were screened for size and the good ones were conveyed into a feeder system that eventually blew them out to the waiting railcars. Chips that were too large were sent to a re-chipper and the small particles, or fines, were conveyed to the mill's beehive burner.

"It took about three or four months to get all the equipment in," says Smith. "We had to pour cement and add onto the building and still keep the mill running around it until the new equipment was ready to go. There wasn't a lot of money around but Dave had a knack of finding old material, steel and whatnot, and he'd put it together so it worked. You had to make do and scrape together and get the parts wherever you could."

As expected, selling chips provided a much-needed source of revenue at a crucial time in the company's development.

Canadian Forest Products started off by paying around $9 a bone-dry unit (2,400 pounds). Ray Williston had convinced pulp producers to agree to a minimum rate. Each railcar held between 28 and 35 units, and each shift was able to produce at least one and a half railcars of chips. Depending on the number of shifts and the size of railcars, Ainsworth was now bringing in up to $20,000 more a month from chips alone.

Within several years the area's sawmills were getting more than $10 a bone-dry unit as rates began to be tied to the price of pulp.

As welcome as the additional revenue was, this new responsibility of producing chips didn't come without initial problems.

Fibreco's chip facilities in Vancouver were soon filling freighters from around the world.

First of all, the product had to meet certain standards or there was no payment. Adjusting the chipping equipment and screens for the correct size of chip took weeks of experimentation.

Furthermore, Ainsworth's contract with Canadian Forest Products stipulated that the chips had to be sorted according to species: pine and fir, and cedar if there was any. Pre-sorting proved to be the wisest course of action for the pulp mills, but it meant that log loaders in the bush now had to avoid mixing species in a single load, and log sorters had to pay more attention in the log yard.

The chips also had to be clean, with no sign of mud or debris in the loads, so blowing the chips into the railcars was the most efficient method of handling. Ainsworth's earliest setup for blowing the chips, however, was rudimentary at best. Initially there was no convenient off-on switch on the blower. If an empty railcar wasn't winched into place in time, the chips simply poured out onto the ground — both a mess and a waste.

The blower was a bit unwieldy, and using different-sized freight cars that were not designed for transporting chips didn't make things easier. When a car was nearly full, an employee had to jump in and rake the chips into the corners to ensure a full load.

Within a year or two the technology had improved considerably. David paid a visit to his friend Dave Hopkins in Merritt, who had developed an adjustable blower spout better suited to loading cars of various shapes and sizes.

After seeing the equipment in action, David returned to 100 Mile and quickly set about building his own version of it.

"This new rig for blowing the chips probably cost us $50,000 but saved us hundreds of hours of fiddling around," he says. "These things became a necessity and it got down to a science where it worked quite well. Every one of our cars would be heaped up from end to end, nicely loaded and crowned, so that when the train started moving the chips would jostle and settle down.

"I'd see other mills with a string of cars but they were never filled. I'd drive by the Exeter rail line and smile to myself because we'd taught our guys to do a better job of filling them."

Meanwhile, if railcars were unavailable, the chips were blown into a pile in the yard. Then a front-end loader with an oversized bucket — used only for loading chips — would be employed to fill the railcars. Not only was this more tedious, but debris could be mixed in with the chips. In the initial stages of Exeter's chip production, an employee inadvertently loaded a large piece of scrap metal from an old conveyor into a chip car.

"One of the guys had been loading chips, and I guess he had dug right into the bank with the loader bucket," says David. "It cost the pulp mill a lot in damage and money. They were not pleased. I said, 'Okay, we'll do something about it. The best I can promise you is that that guy won't be loading cars anymore. I or Allen or Brian will do it and we'll make damn sure we're not digging into the bank.' And we never had any more trouble after that."

Technology for chip production would continue to evolve over the years, and by 1977 Ainsworth would be one of 25 Interior producers to form a chip export consortium called Fibreco.

In the meantime, the extra revenue was a lifesaver. "The barking and chipping gave us a real boost, it got us off the ground," says Al Smith. "We had more product and more revenue, we cut our unit costs, ran steadier, got more wood through, and we didn't damage the equipment as much."

The years since moving up to the Exeter site had seen a steady progression of change and growth. There was far more to come, though, as mechanization continued to take hold in just about every aspect of the industry.

For Ainsworth to remain in the game, standing still simply wasn't an option.

Susan and David at an industry gathering in the 1960s.

Chapter 5

AGAINST THE GRAIN

The concrete foundation has been poured for a major upgrade at the Exeter site in 1970.

By the late 1960s the Cariboo's hundreds of bush mills had disappeared, and many of the larger, stationary mills were entering a new phase of productivity and technical innovation.

As an example of the industry's restructuring, the number of operating mills in B.C. actually declined from 1,627 to 627 between 1962 and 1971. At the same time, lumber production increased nearly 50 per cent.

The markets were relatively strong, the demand for chips was bringing in additional revenue, and new sawmill equipment was revolutionizing lumber production throughout the region and beyond. High-speed band saws with thinner, more accurate cutting blades played an important role, but a new machine called the Chip-N-Saw was at the centre of the changes.

The Chip-N-Saw was originally designed by Ernie Runyon, an electrician at a sawmill in Shelton, Washington. A Vancouver manufacturer, Canadian Car Company (CanCar), took over his design and built the first production model in 1963. John Ernst, a Quesnel newspaper publisher and owner of Ernst Lumber, was the first buyer of a Chip-N-Saw.

The Chip-N-Saw offered an entirely different approach to processing logs by using a linked series of chipping blades and circular saws in a single unit. Once a debarked log went through the machine, the saw's three sets of chipping blades reduced it to a four-sided cant. The cant would then continue down the line, where two sets of circular blades would cut it into two-inch boards. The boards were further cut into 2×4s, 2×6s or 2×10s. The chips, meanwhile, were sold for pulp production.

The beauty of the Chip-N-Saw was its high efficiency and reduction of waste.

Sawdust production in some mills dropped from 25 per cent down to 7 per cent. Thanks to this new machine, roughly 70 per cent of the log could now be processed into lumber and the remainder could be sold as chips.

The Chip-N-Saw also contributed to major changes in the way sawmills were designed, based on a new principle of continuous flow. Traditional carriage mills moved a log back and forth through a fixed saw blade to cut the log into lumber. When one log was finished, the process was repeated. Milling small-diameter logs took almost as long as larger-diameter logs. Overall, the entire process was very inefficient.

The Chip-N-Saws could handle logs on a continuous basis at a rate of 90 feet per minute, compared to about 20 feet per minute on a gang saw. Within several years the new saw, outfitted with more powerful motors, could handle wood at the rate of 300 feet per minute. The continuous forward movement of the logs meant that even small-diameter pieces could now be milled efficiently into dimension lumber, predominantly 2×4s and 2×6s.

This increase in log-processing speed made changes necessary up and down the production line. Faster debarkers, saws and lumber-sorting machinery soon came on stream in an effort to keep pace.

The Ainsworth mill, meanwhile, was relying on 1950s technology and was quickly losing ground. The average diameter of logs brought in from the bush was steadily decreasing, and the outdated equipment made it difficult to adapt.

The Marathon two-saw scragg still relied on 50-inch circular saw blades that had a three-eighth-inch kerf. (The thickness of the saw cut is referred to as "kerf." The greater the cut's kerf, the more waste is produced in the form of sawdust.) Past the twin-saw, the bull edger used up to 10 or 12 saws that also had a three-eighth-inch kerf.

"We would produce a mountain of sawdust in a shift," says Allen Ainsworth. "And the conveyors just couldn't take it away. We would add flights and more flights and then put sides on the conveyor so they were three feet deep, and the sawdust poured into the burner like a waterfall. Within two hours of starting the mill and having a good run, we would plug the burner. Then the material would lie against the side of the burner and burn the sides out. We literally couldn't burn all the waste we were producing."

Given the mill's existing technology, the recovery rate from log to lumber was less than 50 per cent. Knowing that superior equipment was now available and being put to use in the region, the Ainsworths decided in 1969 to build a modern new stud mill at the Exeter site.

> *Pouring cement after the frost came wasn't unusual at Ainsworth, and one time we were pouring foundations for more dry kilns. Ainsworth always had more dry kilns than anyone else, primarily from using hot water to dry the lumber. It was slower but it did a better job. We had all the foundation bolts in, and it was a rather intricate foundation with a mass of bolts down one side to hold the panels. That night we had about six inches of snow, and the next morning one rather dedicated employee decided he would help us clear all the snow off the footings. He was using the bucket on a 950 loader. He just ran the bucket across the top of the footings and, of course, he sheared off all of our anchor bolts that we had put in the freshly poured concrete the day before. We had to redrill the concrete and put them all back in again. Fortunately, it was late in the year and the concrete wasn't really set all that well.*
>
> — KEN GREENALL

Exeter site in mid-1960s.

Al Smith's sawmill expertise played a key role in the company's growth over many decades.

The Ainsworth office was a no-frills affair in 1969.

Ken Sato at the weigh scales as another load of logs arrives at the Exeter site.

Steel work for the Exeter site's new stud mill in 1970.

The creative challenge of constructing a brand-new facility, along with the prospect of greatly improved recovery and a huge jump in production, provided ample motivation to get started.

The mill was nearly a year in the planning and would be built around two Chip-N-Saws. Most logs that went through the small Chip-N-Saw would be processed into 1×4s, 2×4s and 2×3s and sent down the sorting chain to the lumber stacker before being transported to the drying kilns.

Other small logs that required re-edging and resawing would go to a thin-kerf double arbor edger before going to the stacker and on to the kilns.

The bigger Chip-N-Saw would process logs up to 20 inches in diameter and cut them into three- or four-inch cants, and later six-inch cants. The cants would then be turned over on their sides and re-edged with a twin band saw into various dimensions of lumber. Others were sent through a double arbor resaw for further processing.

The Chip-N-Saw units cost nearly $250,000 each, and the new mill's total cost was approximately $4 million. Financing would come from a combination of banks and the Industrial Development Bank.

"It was a big, awkward step for us to take, but we had to get into higher recovery," says David. "At the same time, we were already getting into the smaller-diameter logs."

The design for the new mill and the output projections looked good on paper, but the ensuing couple of years proved to be among the toughest that Ainsworth would endure. Some of the problems, especially those revolving around scheduling estimates, were self-inflicted. Other setbacks were the result of technical issues beyond the company's control.

During the initial phases of construction the original stud mill was kept running as long as possible to continue generating revenue. Eventually the old mill had to be shut down completely and dismantled to make way for the new building. It was sold for cents on the dollar to Bond Bros., who transported it north to the Fort St. James area.

This was an extremely critical period. With production limited to a small number of Stan Halcro's cants going through the gang saw, revenue dropped precipitously — all the more reason to get the new mill up and running as quickly as possible.

CanCar, the Chip-N-Saw's manufacturer at the time, didn't help matters. Their revolutionary equipment was in such demand that they couldn't keep up with the orders. Of the two models purchased by Ainsworth, the Chip-N-Saw dedicated to large-diameter logs was three weeks late on delivery. This in turn held up construction and equipment installation since so much of the mill's design hinged on the saw's placement.

Most of Ainsworth's 90 or so employees were put to work at the site. Among those who played an important role with their construction or steel-fabricating skills were Bob Watson, Al Smith, Bob Glover, John Classen, Charlie Kraus, truck driver Dick Dickson, Danny Snowden, Stu Marshall, Lyle Jones, Vern and Clarence McMillan, and Bill Becker.

Allen Ainsworth dug out the foundations with a small backhoe, and the cement was mixed in a one-cubic-yard mixer purchased from the City of Vancouver.

Brian Ainsworth and Garry Babcock were brought in from the bush and put to work fabricating the pieces for an as-yet-untested lumber-sorting and-stacking system devised by David. Known as a J-bar system, it was a design he had always wanted to try out. Now he had an opportunity.

Being able to use so much in-house help, as well as the Ainsworth knack for improvising when resources were scarce, made a sizeable difference in the Exeter mill's construction costs.

On one occasion, Bob Glover and several others returned to the mill after hours with Allen Ainsworth to do some heavy lifting, which was better done when the site was empty and free of distractions.

"For safety reasons, we had gone back in the evening to use a loader to lift the debarkers onto their mounts," says Glover. "It was a little precarious because we had a 966 loader lifting the debarker and a 950 loader chained on behind the 966 just to hold its back end down."

They succeeded in lifting the two huge Cambio debarkers into place, and when the other employees returned to work the following morning, they wondered how the equipment was ever moved.

"Nowadays you'd use a crane, but back then we did what we could with what we had," says Glover.

> *People used to run down a steep bank and come flying in the back door of the old shop. So we thought we'd announce them as they were coming in. We took an air horn off a machine we were working on and rigged it up so that it went off as the door opened. Of course we had it pointing straight down for the best effect, and we got all kinds of reaction.*
>
> *Some people dropped to their knees and just about had a heart attack. Other people turned around and went right back out the door. Pete Nadin came through like nothing had even happened. It was only there for a day or two and then we took it down. We probably needed to put the horn back on the vehicle.*
>
> — BOB GLOVER

Once the bigger Chip-N-Saw arrived, transporting it into the mill made the debarker installation look easy. Other pieces of equipment had already arrived and been moved inside as close to their final positions as possible. The Chip-N-Saw components had to be moved into place through a gaping hole left in the wall.

"It was a helluva job," says David Ainsworth. "The forks on the loader weren't suitable to lift the Chip-N-Saw, so we had to improvise and finally weld it right onto the loader's fork frame. Then we had to add weight on the back of the loader, which Allen was driving, so about a dozen of us were hanging off the back, and we just staggered it in there. It was quite a performance. We were pretty desperate to get it going. Then we worked night and day to get it ready and hook up the electricals."

By this time the bills were starting to pile up and the banks were growing increasingly vocal about their concerns. With pressure mounting to begin producing lumber, the situation went from bad to worse: the larger of the two Chip-N-Saws was continually jamming up because of a mechanical defect. This delayed the mill's startup even further.

It seems the CanCar shop was so busy churning out their new product during this period that quality control went by the wayside.

"It was really a disaster; we had colossal problems with it," says David. "They sold too many too quickly. They tried to put production on three shifts so they had a lot of new people who made a few mistakes and the customer had to pay for it."

Worse yet, the CanCar personnel sent up for the saw's installation and troubleshooting were ineffective. No sooner would they arrive on site than they would be called off to solve

Exeter mill site in the mid-1970s.

Chip-N-Saw problems at Sam Ketcham's West Fraser operation, which had purchased four or five of the new units.

It was Ainsworth employees who eventually determined that a feed roller designed to follow the contour of the log was welded into place incorrectly. This caused the machine to jam up when it hit the thicker part of a log, often burning out a motor in the process. Since there was no reverse on the machine, these half-chipped cants wedged into the bowels of the saw had to be winched out with great difficulty. And the burned-out motors had to be replaced.

Nearly two weeks passed after delivery with not a single cant making it through the new Chip-N-Saw. Half-cut logs were strewn about the mill floor.

"They had welded this roller on there like a bunch of dummies and it took some of our guys to notice it, so I had to come along and say, 'You're moving that goddamned roller and that's it,'" says David. We weren't going to put up with this any longer. We finally had to say, 'Get the hell out of our way and we'll do it ourselves.'

"We had all of these problems around our ears and they were just choking us to death. We had over-extended ourselves so badly by this sort of thing. It doesn't take long to lose thousands of dollars while you're fooling around with someone who doesn't know what he's doing."

Unfortunately, startup problems weren't limited to the new Chip-N-Saw. Despite their best efforts at planning, many of Ainsworth's conveyor systems had to be rebuilt and redirected to keep the wood, chips and waste material flowing properly.

Most established mills of the day experienced similar problems due to the revolutionary nature of the new equipment. It took Interior sawmills months, or even years, to integrate all the latest machinery and ramp it up to full production.

Production was ramped up in the bush to provide the new mill with a sufficient supply of logs.

The learning curve for mill workers was steep as well, especially on the Chip-N-Saws. This marked the first time employees sat in a glass-enclosed booth and controlled the flow of logs with a dizzying array of buttons on a panel.

Saws were inadvertently bent and broken and equipment modifications seemed to take forever, but by 1972 the Ainsworth mill was producing lumber on a consistent basis. The chips were flowing to the pulp mills and the benefits of higher log recovery and increased production were being realized.

Monthly production jumped to approximately 10 million board feet, up from 2 million with the old mill. The log count underwent a similar transformation. The old mill was capable of processing between 2,000 and 4,000 logs per shift. With the new Chip-N-Saws now in place, the mill could handle up to 10,000 logs per shift.

"We eventually started doing 11 million board feet a month with that mill," says Allen. "We moved into automated stackers, an automated lumber-sorting system, and we had about 10 people on the floor doing four times as much production.

"The Chip-N-Saws allowed you to have sorted logs — sixes and sevens and eights — and then we'd run 100 logs of one diameter on a particular Chip-N-Saw set, so it was just one long ribbon of logs. The lumber was cut and chips came out the back end. They just poured out. It was perfectly well-sawn lumber, and not bad recovery at all."

Before long, these production increases at the mill precipitated a number of changes on several fronts — in the plant, the sales office and the bush.

More lumber meant that more drying kilns were required, so the site soon had a total of six, along with a second hot water boiler to provide the heat. The boiler system was performing admirably, but there was one significant drawback: the shavings bins that fuelled the boiler had a habit of catching on fire.

"There was a burner by the planer and another burner for the sawmill, so the sparks off the burner would get into the shavings pit and then we had fires," says Bob Glover. "There was no real fire protection. We had to pump water out of a ditch."

> *Dave always said that if you operate a piece of machinery in the company, you should treat it like it was your own because we couldn't afford to replace it. If he saw you treating equipment badly, you were stopped right then and he'd get you out of the machine and you'd get your lecture. I've been to other companies with Dave and he would just shudder to watch how other people drove equipment there. We were in Quebec on a trip years ago and there was a guy driving a loader, just destroying it, and David was having a fit. Allen was saying, "Dad, we don't own that equipment!" But it still bothered him that someone was out there trying to destroy this piece of equipment. He knew what it was worth and he looked after his stuff.*
>
> — AL SMITH

The logical solution was a steel storage bin. After a good deal of trial and error — even experimenting with rope on the shop floor — David developed a design for a round bin that outperformed any other models in use at the time. One of the main features was a revolving circular chain on the floor that kept the shavings loose and directed them to the burner.

"People had built bins like this before, but there was never a system that worked very well for dragging the shavings from underneath and feeding them into the burner," says long-time Ainsworth building contractor Ken Greenall.

"Everybody had grief with them. Dave's design was ingenious and it worked extremely well. I haven't seen one before or since that worked as well as that one."

The construction of the new mill also coincided with changes in how the company's lumber products were marketed and sold.

At the Exeter shop with hauler Doug Cathro, Bob Glover and Len McIntosh.

Up to 1970, Vancouver lumber wholesaler Battle and Houghland handled the sales and marketing for both Ainsworth and the Pinette and Therrien (P&T) operation in Williams Lake.

The principals of Battle and Houghland were looking ahead to retirement, so Ainsworth and P&T joined forces to form their own sales company. They called it Mountain Pine Lumber Ltd. Conrad Pinette, Gabe Pinette's son, was company director, and Allen Ainsworth was secretary.

By 1975, Herb Weier of Weier's Sawmills in Quesnel would join the cooperative venture, making it a threesome.

Two of Battle and Houghland's senior salesmen, Doug Morter and Con Cawley, took new positions with Mountain Pine Lumber at its inception. For Ainsworth and P&T customers, the transition from one company to the next was seamless.

"Con Cawley and Doug Morter were from the old school and they fit right in to the industry," says Allen Ainsworth. "They were pretty forward and aggressive, they drank and smoked and travelled a lot and they worked like horses."

The switch from Battle and Houghland to Mountain Pine marked the beginning of Allen's direct participation in the company's sales function. From now on he'd be travelling to Mountain Pine's North Vancouver offices at least once a month, in addition to taking numerous sales trips further afield.

The new stud mill at Exeter also opened the door for Tom Ainsworth to rejoin his younger brother in the Cariboo. Since leaving the partnership with David in 1952, Tom had continued logging in the Harrison Lake area near Chilliwack.

"I kept saying to him, 'You know, we have to find something for you to do,'" says David, "because I was concerned about the danger of logging. He had been there long enough.

Mechanized logging began with snippers on a loader.

Early machines were slow because they had to approach each tree individually.

I finally convinced him that he should come back up. He didn't have a financial interest in the business anymore, but he was a very capable person and we needed people to do the saw filing, so I convinced him he should do that."

Tom learned the trade by visiting other filers, watching them work and asking questions.

"Lots of filers could be really miserable about sharing their information but Tom was an engaging sort of person," says David. "He was interested and knew quite a bit already, so he'd go in and talk to a filer for a while. Then quite often the guy would break down and tell him things — the little tricks of the trade he's got. It's a very skilled occupation, a real science."

Modern stud mills that could process up to 10,000 logs per shift had a major impact on logging methods in the bush. Whereas 6 to 10 loads a day used to be sufficient to feed the scragg mill, a daily diet of 20 to 30 loads or more was now the norm. Virtually overnight, there was a need to increase production.

Bob McCormack with the Drott 40, the first feller buncher.

Fortunately, mechanized falling machinery was beginning to come on stream in the Interior, but the earliest versions of the equipment were barely up to the job. Compounding the problem was the fact that the Ainsworths and their early contractors had little opportunity to compare notes with other operators. Most in the area were still logging large-diameter fir trees; with fewer pieces to handle, other operators felt less pressure to ramp up production.

Both Ainsworth and its contractors had a mutual interest in doing whatever was necessary to make the equipment work and deliver logs to the mill. This involved a high level of trust and cooperation. In many cases, Ainsworth's business relationship with its logging contractors went on to span several decades, despite numerous challenges along the way. If equipment didn't cause problems, then weather conditions often played havoc with logging and hauling schedules, while unfavourable market conditions could lead to production curtailments or work slowdowns.

Mechanized logging equipment, however, was an absolute necessity.

"Some of these early machines were just terrible but you had to use them," says Brian Ainsworth. "There was nobody there to help you, and not getting it done was not an option. You can't make lumber out of excuses."

Whether it was feller bunchers for cutting the trees, grapple skidders for moving them through the bush, or delimbing equipment to remove the branches, the equipment was being modified almost constantly, regardless of whether it was owned and operated by Ainsworth or by its contractors. The coming years would see a rapid succession of improvements in both performance and durability.

Early on, however, the new owners of this equipment were largely on their own, especially in the far-flung reaches of B.C.'s Interior. It was up to talented Ainsworth employees like Bob Glover, Len Blomfeldt, Don Fraser and Grant Dohman, as well as the enterprising logging contractors, to keep these machines up and running.

After performing repeated repairs and modifications, an Ainsworth mechanic often knew more about a piece of equipment than the manufacturer's representative visiting the site.

Contractors did their share of monkeywrenching as well.

Larry L'Heureux, who started logging for Ainsworth in the early 1970s, spent so much time repairing early stroke delimbers that he would sleep in his truck at night for several days running rather than waste precious time commuting back and forth.

Crawler tractors or basic loaders outfitted with hydraulic shears were among the first pieces of equipment designed to replace hand-falling with chainsaws. The Cat or tractor had to be driven right up to each tree, however, which made it impractical for stands of small-diameter lodgepole pine. Its lack of maneuverability in the bush was also a problem.

More useful — at least in theory — was a relatively new piece of equipment called the Drott 40 feller buncher, designed by Wisconsin equipment manufacturer Erv Drott. This machine was also equipped with a hydraulic shear, but the shear was positioned at the end of a moveable boom, along with a grapple to grab the log and set it down in a pile.

When it worked properly, the Drott could snip and pile up to 100 trees an hour. Early versions of the Drott were plagued with problems, however.

"We were starting off with equipment that wasn't purpose-built for the bush," says Brian Ainsworth. "The Drott was basically a backhoe or excavator that was meant to dig ditches on level ground, and now you're trying to crawl over stumps and go up hills and climb over trees with it. When it was meant to push, you were pulling, and when it was meant to pull, you were pushing."

> *We started all sorts of schemes along the way to get to roadside logging. Every time a new piece of equipment came out, it was going to be the answer to everything but it never seemed to work out that way. Then someone else comes out with something new and it starts all over. Those first stroke delimbers, they said they would work 24 hours a day, seven days a week, but they broke down a lot.*
>
> *Brian Ainsworth was always showing up some place, usually when you didn't want to see him — like any boss I guess. He'd show up at the goddamnedest times. But he always wanted to improve things. You could talk to David or Brian and say you could do things this way or that way, and they'd say, "Well how much is it going to cost?" You told them and they usually went along with it, and then you hoped it was all going to work better.*
>
> — LARRY L'HEUREUX

Snipping the trees off with hydraulic shears was faster than hand-falling, but it also shattered the log five or six feet up the trunk in cold weather, making that portion unusable for lumber. As for its reliability, operators spent as much time lying underneath the Drott in a pool of red oil replacing final drives and brakes as they did sitting in the cab and cutting down trees.

Given their importance in providing a steady flow of cut timber, feller bunchers underwent numerous improvements with each passing season. Increasing numbers of manufacturers were also entering the market as mechanized logging became the industry norm.

Brian Ainsworth directs a Cat operator with a full load on the Hayes logging truck.

The feller buncher's hydraulic shears were soon replaced with chainsaw blades, which in turn were replaced by a circular cone saw to cut the tree. Heavier steel was used throughout, lifting power was increased and downtime was reduced. Over time the machines became far more maneuverable in the bush under a broad range of conditions and terrain, especially when rubber-tired models became available.

With feller bunchers now able to pile trees uniformly at the butt, skidders with large grapple attachments could be used to grab the bunched logs for transport to a landing for processing.

Early choker skidders first ran on crawler tracks, then glorified tractor tires. Now they were being replaced by powerful large-tired grapple skidders that could traverse stumps and drag logs over steep terrain or through six feet of snow.

Garry Babcock, who joined the company in 1967, became Ainsworth's first skidding foreman not long after he suggested making a switch from piecework to hourly rates for the skidder operators.

"We used to get 13 cents a tree on the landing," says Babcock, who started out on a C4 Tree Farmer skidder. "And then you had downtime and you did your own monkeywrenching for $2.50 an hour. Some guys were putting in for 200 trees a day, and I had a real struggle skidding anywhere near 200 trees a day. So I was wondering how these guys were doing it. Well, they were padding the count. Susan Ainsworth picked this up in the office because the amount of lumber produced for the amount of trees coming in didn't jive. She picked it up on the tally."

The early stroke delimbers required constant maintenance to keep them operating.

The next step in the process of getting trees from the bush to the mill was removing the tree's limbs and topping it before it was loaded on a truck.

"Limbing was one of the biggest hurdles," says Brian Ainsworth. "With only six to eight loads a day, guys could get out there and whack them off by hand with chainsaws. Once you're into 20 to 30 loads a day you can't do that."

Before the advent of the mechanized stroke delimbers, flail delimbers were used at the landing where trucks came in to load up. A separate power plant or engine drove a drum with chains that would simply break off the limbs.

"With the chain flail you had to handle the tree twice," says Brian. "You would chain flail it, pick it up and roll it over, and then you had to go at it with the chainsaw for quite a while to get it to a state where you could bring it to the mill. It was crude, but compared to cutting by hand it was okay."

The shift to mechanized logging methods came with a hefty price tag.

Ainsworth's first Drott 40 feller buncher cost $75,000, and as equipment became more sophisticated, the prices rose accordingly. Individual pieces were ringing in at between $250,000 and $400,000 within 10 to 15 years.

If the Ainsworth mill required 30 loads of logs a day, they needed contractors who had the right equipment to do the job. In some cases this meant assisting contractors with financing, especially in the early days of mechanization when fledgling operators couldn't secure sufficient credit on their own.

More often than not, the company was paying too much for its logs, but the compensation was necessary if contractors were to be enticed to invest in new equipment.

"When it came to the equipment, you had to have an operator who knew what the machine would do and then didn't break it by going too far," says Brian. "The machine sitting there broken doesn't help. If you don't get things together and you don't get the loads in there, nobody is going to help you pay for anything; there's no big pool of money to work with.

"We had good guys, though. They were all keen and helpful and wanted to get it done, and we did it."

Ainsworth logging contractors such as Larry L'Heureux, Dowling Monette, Don Barrick, Marvin and Jim Higgins, Gordon and Merrill McNabb, and Gerry Blais all played an important role in the transition to mechanized logging methods.

The gradual introduction of heavy equipment in the bush didn't mean that the challenge of supplying these high-speed mills with a steady supply of small-diameter logs had been solved completely.

The issue remained of how best to take advantage of the mechanization, which meant developing a method of logging and handling the small wood that was both efficient and cost-effective.

The traditional method of skidding the logs to a single landing in the bush where they were limbed, bucked and loaded onto trucks became unworkable as the number of pieces steadily increased. Forestry regulations were now limiting the size of landings, so space was at a premium.

Huge piles of limbs and tops crowded the landing, making it difficult for the skidders, loaders and logging trucks to maneuver safely. Bucking crews were especially vulnerable in this busy environment, and the advent of flail machines by the mid-1970s added to the danger.

The other drawback to the status quo was an over-reliance on one piece of equipment. If the feller buncher broke down or a skidder operator failed to show, the day's production could grind to a halt. Each phase was linked to the other.

> *When the first Drott feller bunchers came out we were always losing the final drives on them. Dave said that to warrant these new machines we had to run them in two shifts. If we lost a final drive on the night shift, it was my baby, so they'd throw the final drive in the back of my pickup and I'd head for Prince George. The deal was they would send someone down from Prince George with a replacement and we'd meet halfway in McLeese Lake for the handoff.*
>
> *One time it's about 2:30 in the morning and I'm waiting in McLeese Lake for these two turkeys who were supposed to be coming down from Prince George. They didn't show up so I just kept on driving north because I have to have a final drive. Our machine is down. So I get as far north as Quesnel and I spot their pickup at a motel. They've got two honeys in the motel, and I spot my final drive in the back of their truck. So I get them up and we put that final drive in the back of my truck, and I give them the one I had. These guys were real characters and they didn't last long. The dealer got rid of them.*
>
> — GARRY BABCOCK

It took a full 10 years before a suitable system of logging and handling the wood was devised, and there was plenty of trial and error along the way.

"The transition was really tough," says Brian, "but it did evolve. You had to keep trying different things because what used to work didn't anymore. It was just the sheer number of pieces."

One of the costliest experiments along the way involved a massive landing or reload station, not unlike a major airport's airstrip in size, that was carved out of a patch of bush. The goal was to de-link the various phases of logging, thereby allowing work to continue if one phase was held up.

The logs were skidded from the bush to a roadside or landing where they were loaded onto logging trucks — limbs and all. These whole trees were then trucked to the reload landing, spread out and eventually limbed and topped and sorted.

De-linking the various phases in logging was the key to boosting production.

Once ready for the mill, the logs would be loaded onto logging trucks again and hauled into the Exeter site.

Although the concept looked promising on paper, the end result didn't work as intended. For one thing, the reload landings were costly to build because of their size. Wet weather turned them into instant bogs because the cost of gravelling them was out of the question. The number of trucks coming and going was a nuisance, and the huge piles of debris, which had to be burned virtually non-stop to keep up, were getting in the way.

In addition, forestry officials warned against the size of the landings.

"Going into this, we thought it would work," says Robin Nadin, the logging foreman working directly under Brian Ainsworth. "Of course, with all that extra handling it didn't work, but something always came out of something that we did wrong. There was a whole evolution of changes there. Brian would have to go back and answer for this, but he always said it was easier to beg for forgiveness than to ask for permission."

It wasn't until the early 1980s that Brian Ainsworth and contractor Larry L'Heureux settled on a system called roadside logging, which has now become standard practice in the industry.

Although roadside logging was beginning to be used in parts of Ontario and near Grande Prairie in Alberta, Ainsworth and L'Heureux can safely be credited with pioneering the practice for application in the Interior.

The advantages of this new system were huge — for the company and the contractors.

Skidders now dragged bunches of logs out of the bush and strung them out butt-first alongside a logging road, where they were processed and loaded onto trucks.

The steady introduction of improved equipment over the years played a large part in the system's development. The stroke delimber had just been introduced, and although early models were notorious for their downtime, they replaced the bucking crews and chain flails.

The introduction of the butt 'n top loaders made roadside logging possible.

The entire operation was now fully mechanized. Employing two shifts on the delimber, Larry L'Heureux could keep up with any amount of logs the grapple skidders hauled to the roadside.

A relatively new type of loader called the butt 'n top was also instrumental in the development of roadside logging. Ainsworth happened to have two Barko 450 butt 'n tops sitting idle, left over from the ill-fated reload landing experiment. Thanks to the butt 'n top, it was now possible to load a logging truck on uneven ground, or at a roadside, rather than at a level landing in the bush.

No longer tied to a congested landing, each phase of the logging operation could progress independently. It meant equipment operators could stay out of each other's way. If necessary, large inventories of logs could be stored at each point in the process until the skidders or delimbers showed up to do their jobs.

There were significant gains in productivity, and the cost of logging decreased accordingly.

"Larry L'Heureux was the first guy who was successful in bringing in the new equipment and figuring out how to get all these things working together," says Brian Ainsworth. "The falling, the bunching, the skidding and two shifts on the stroke delimber — it was ingenious."

By now a new form of timber licence — a Timber Sale Harvesting Licence — had been

Using excavators to build logging roads was a new development in the Cariboo.

Logs were still used in bridge building prior to today's modular units.

introduced. It was created to provide a longer-term solution for licensees who wanted cutting rights on the region's Crown land, and required the submission of a five-year forest development plan.

Meanwhile, the combination of the feller buncher and the grapple skidder was far more suited to clearcut logging than to selective logging, where only certain trees are removed. As a result, government forestry regulations increased in scope and number in an effort to keep pace with mechanized logging and the changes it fostered.

Lumber producers now found it necessary to have a forester on staff to look after everything from securing new sources of timber to ensuring that the province's silviculture guidelines were being followed after a cutblock was harvested.

Pete Nadin became the company's one-man forestry department in 1971 after an 11-year stint with the B.C. Forest Service and several years as a logging and bucking contractor.

Gone were the early bush mill days when the forestry regulations covered little more than stump sizes and ruts in the skidding path. Nadin's newly created position required him to acquire cutting rights and submit logging plans that detailed the size of landings, location of roads, logging methods and even how the landings would be rehabilitated for grazing purposes.

With several government ministries involved, it could take up to a year for forestry plans to be approved. In addition, more and more of the planning and preparatory work was falling to the licensees.

Operators were now branching out into more remote areas, which meant that the construction of logging roads had to follow suit. Unlike the majority of mill owners at the time, David Ainsworth felt there were distinct advantages to having the company build and maintain its own logging roads rather than handing the job off to contractors.

His reasoning was straightforward: the upfront equipment and construction costs for the roads were substantial — well over $16,000 per mile at the time — but they would be offset by future savings in reduced maintenance costs and less wear and tear on the vehicles that used them.

The first time I met Dave Ainsworth was not too long after I went to work for him. I was backing up my truck out of the shop and I had my logging trailer hitched on. I was backing over the pit in the shop and looking after the trailer so it didn't jackknife. I forgot about my front wheel and put the truck into the front of the pit. It was on a Saturday and the only one I could find to give me a hand to get it out of there was Dave. That was one of the most embarrassing moments.

Another time I was coming out of Spout Lake during the winter. I still don't know what happened, but I looked out my mirror and there was my trailer, upside down. I started walking back to the landing to get a loader and after I had gone three or four miles I saw a guy on a Cat. I went to ask him to help and by gosh it was Dave driving the Cat way out there in the bush. That was the second time I got into trouble and had to have him help me out of it.

One other time I was driving a scraper in the log yard during breakup because we weren't hauling logs. There was a big hole in the middle of the yard and I drove right into it and got really stuck. Dave was around again to pull me out. This time he told me a little story that ended up, "If you can't get out on your own, for god's sake don't go in!"

— DICK DICKSON

Whether the trucks were owned by Ainsworth or operated by contractors, the quality of the logging road had an impact on their overall performance.

Better-constructed roads also provided for additional days of hauling during breakup and bad weather when other roads were too soft to drive on.

As it turned out, Pete Nadin's brother Robin took charge of the company's road-building activities after 15 years of bucking and logging for Ainsworth.

"The advantage to building our own roads was that we were totally in control of what type of road we were getting," says Robin Nadin. "David maintained that you're going to pay for that road one way or the other — either when you build it or when you maintain it. And you're going to pay for the trucks when they break because you're busting them up going over a bad road.

"When I first started in road construction, David used to beat on me because he was more of a grader person. Of course, the last thing you do on a new road is grade it up, and if you've got any flaws, the grader is going to find them. David would say, we've got a hump here we shouldn't have, and so on. The crown of the road was his big thing, and he was right. He told me that if you control the water, you control the road. If you build it right the first time and put the right base down, it should hold up forever."

The majority of the early road-excavating work was carried out by Joe Danczak on his D4 Cat, with David often finishing up on a Galleon 60 grader, the company's first. In particularly wet areas where long-term stability was an issue, it wasn't uncommon to push out the mud and lay down a base bed of logs, corduroy-style, before piling on the dirt.

> *I guess I always wanted to build roads. My dad said that when I was four years old I wore out 17 pairs of jeans and four 2x4s just building roads. We never had Dinky Toys or anything like that; I had a piece of 2x4 with a shingle tacked on the front for the blade, and that was my Cat. I never had a bought truck until I was about 11, and somebody left that at our resort. It was never bought for me. Dad used to say, "If you're looking for Robin, he's out in the dirt pile, building roads."*
>
> — ROBIN NADIN

Contractor Larry L'Heureux was instrumental in developing roadside logging.

Susan Ainsworth had a surprise in store for Garry Babcock.

I'd had a bad day in the bush. It was cold and freezing and I wasn't in a very good mood when I got into the office. Dave is there and he says we've got to go down and put some salt on the kiln tracks because everything is freezing up. That's the last thing I want to do. I'm fed up after a bad day and this really puts the cork in the bottle. So I guess I start swearing about what a so-and-so Mickey Mouse outfit this is, and apparently Susan heard me.

Later that year at the Christmas party, Dave is giving out the 10-year watches, and when it comes to me, Susan gets up to give me my watch and I think, this is a bit different. So I get up to the front and she gives me my watch, and it's a Mickey Mouse watch. That's what she gives me. My mouth falls open and she says, "Well, you work for a Mickey Mouse outfit, don't you?"

And right away I thought, "Oh no, she never forgot about that." After we had a good laugh, I also got a beautiful Seiko watch.

— Garry Babcock

As forestry regulations grew more stringent in response to environmental concerns, the guidelines for road and bridge construction also came under review.

"Early on we used to be able to put in a simple bridge in under two hours," says Joe Danczak. "Just using a Cat with a logging arch on the back, I'd put in two big logs on either side, and then they'd fell some smaller trees for the decking."

Within a number of years, bridges had to be designed by professional engineers and featured wood decking bolted to steel girders. Similarly, punching in a road with a D8 Cat was no longer acceptable, and excavators or backhoes were now required for the job.

Road building with excavators was well established at the coast, but it was new to the Cariboo. The new road's stumps and debris were actually left on the grade, which was then covered with material taken from both sides of the grade. This technique left ditches on both sides of the road, with the road elevated up to four feet above the surrounding terrain. The result was better drainage, with less maintenance required over the long run.

Doug White and another woodlands staffer set out to survey the area's timber supply.

In 1972, British Columbia's Interior sawmills were among the most productive and efficient in North America, if not the world. A building boom across the continent was driving lumber prices to record highs, and the volume of wood logged in the province was three times greater than just 15 years earlier.

By 1973 the new Exeter stud mill was in full production. The lumber market, fickle as ever, was now on its way down, although chip prices were up.

It was around this time that the provincial forests ministry announced a timber sale or opportunity licence for approximately 100,000 cubic metres (the unit used rather than cubic feet or cunits with Canada's adoption of the metric system in 1970) of lodgepole pine in the Clinton area, about 45 miles south of 100 Mile House. There were strings attached, however. The successful bidder would also have to commit to building a sawmill. This was only the fourth time in B.C. history that such a requirement was made.

This offering of pine wasn't something the Ainsworths had been pushing for. But when the opportunity arose, they decided to go after it.

As David Ainsworth commented in a news report at the time: "I know we've stuck our necks out, but you can't stand still in this business or you're dead. I guess I feel a bit like the mother of the bride at the wedding: I don't know whether to laugh or cry — but we feel the project has growth potential."

For the forests ministry to designate the pine as timber quota wood rather than a weed species was still a relatively new practice. The NDP government of the day was also looking for ways to create much-needed jobs in the Clinton area. A new sawmill using uncommitted pine from the surrounding region was as good a prospect as any.

Before Ainsworth could submit a bid, the company was responsible for cruising the timber and gathering up the necessary data required by the ministry. The first advertisement for the bids appeared in March, with a deadline of May 17. Early spring with several feet of snow still on the ground was a poor time of year to be cruising the bush, but there was no time to waste if the deadline was to be met. Even when the cutoff date was extended to July 19, it was still a tight race.

There were over 100 enquiries on the timber sale, but when David submitted the company's bid in Williams Lake five minutes before the eleven o'clock deadline, he was the sole bidder. Forests ministry staff reported there were representatives from two other lumber companies looking on at the time.

Despite the seeming lack of interest from other licensees, there was some initial resistance from the local Weldwood operation. The small-diameter pine was of no use to Weldwood's sawmill in 100 Mile, but they saw potential value in the production of wood chips.

"Dad could see the opportunity in the additional volume this licence would give us, in spite of the fact that it was probably some of the lowest-quality timber in the Cariboo," says Allen Ainsworth. "A lot of it was out in that dry-belt area west of Chasm.

"The fact is, he was very forward-thinking and experienced in the harvest and the manufacture of this small wood. He had to be. By this time we were using lots of six-inch, five-inch, four-inch and even three-inch tops. We didn't have any other timber so we pioneered sawing this super-small stuff that Weldwood never wanted. They were getting most of their lodgepole pine from out Canim Lake way."

When the time came to submit the bid, David wrote a cheque for $830,000 as the deposit. At that point it was the highest sum ever required for a timber sale in B.C., which David found incomprehensible — and maddening.

"I went in to see the Minister of Forests at the time, Bob Williams, who was not friendly, and said this was really ridiculous because it was tying up a lot of money that we could dearly use to build the mill," says David. "He eventually came around and we got it reduced a little bit, but it wasn't for another couple of years. In the meantime, that deposit really put a hardship on the project because that money would have gone a long way toward building the mill."

Planning for the permits, road construction and cutblock layouts began immediately. By now Ainsworth's small woodlands department, staffed by Pete Nadin and Bob Day, had doubled in size with the addition of two new recruits: Doug White, fresh out of college, and local resident Bob Phaneuf. White was home visiting his parents at Christmas when Nadin offered him the job, making him the first student in his class at the British Columbia Institute of Technology to land a full-time position.

The mill itself would be built on a 50-acre site formerly occupied by the old Cattermole planer operation, a few miles down the road from a scenic 200-foot-deep gorge called Painted Chasm, after which the mill was named.

For Ainsworth, it was an ambitious proposal.

Based on a timber profile with average log lengths of 16 feet, the plan was to construct a random-length sawmill that could produce 48 to 50 million board feet of lumber annually. The products would include 1-inch, 2×3s, 2×4s and 2×6s with lengths ranging from 8 to 16 feet.

A pressure-treating plant for preserved wood was also incorporated into the design, with the idea of producing treated lumber for foundations.

The various saws and set works for the mill came from equipment manufacturers in Quebec, where small-log processing was more advanced. Rather than relying on Chip-N-Saws, this mill would have a production line of debarkers, chip canters, quad saws and board edgers for greater versatility.

The mill was unusual in that the sawmill, the Stetson-Ross planer and the shipping area were all housed under one roof, which made for a more compact operation.

The initial budget for the mill was $4.5 million, but it cost closer to $6 million by the time the plant was up and running. Even with the aid of various financial institutions, the project quickly ate up any cash reserves the company had managed to set aside.

As with the Exeter stud mill a few years earlier, every attempt was made to do as much work as possible using in-house resources. Work began in earnest on Labour Day in 1974. Contractor Ken Greenall oversaw a crew of 35 men, with Allen Ainsworth and Al Smith taking a hands-on role in everything from designing the mill to pouring concrete and erecting the structure's steel girders.

All of the steel fabrication and design work was carried out at Exco Industries in 100 Mile House. The Ainsworths already had an ownership stake in Exco, which made it the natural choice for supplying all the building components, conveyors and steel for the catwalks. The shop up at the Exeter mill was also pressed into service when the need arose.

It took a year and a half to construct the Chasm sawmill.

Most of the steel work for Chasm was carried out at Exco Industries.

Break time for the Exeter shop crew in 1977. From left: Paul Loeppky, Ed Niemiec, Dave Shaw-MacLaren, Don Fraser, Len McIntosh.

Over the coming months, Al Smith and Allen Ainsworth would truck tons of steel and mill components to the Chasm site. Meanwhile, Ainsworth's transport trucker, Dick Dickson, hauled over 250,000 tons of sawmill equipment from Vancouver to Chasm on his return trips from the coast.

"As usual, there was no money around when we were constructing the mill," says building contractor Ken Greenall. "If you wanted the market to fall, just have Ainsworth start a new project. So we were pouring concrete in winter with not the greatest of materials — the aggregate was difficult to come by — using a one-yard mixer and the front-end loader off the backhoe to load it. Allen and I and one other young fellow stood up all the steel posts and the columns for the whole sawmill. No one else was crazy enough to be doing it, particularly in the green concrete."

The mill took a year and a half to build, with the first logs being processed in February 1976. Doug White and Bob Phaneuf watched the first trees destined for Chasm being logged in the Graham Creek area south of Green Lake.

"Getting the cutting permits and doing the actual logging was very tightly scheduled," says Pete Nadin, "so there was no inventory in the log yard at startup. Those trees were still wiggling when they went into the mill."

Nobody wanted a repeat of the slow and costly startup of the Exeter stud mill in 100 Mile, but the new Chasm operation presented its own set of challenges as the months wore on.

For planerman Carl Watson, the trouble started on day one.

"The startup of the mill had been scheduled and Allen had hired some guy to run the planer, but the day came and the guy was nowhere to be found," recalls Watson. "So Allen drives in to Exeter and says I better head down to Chasm because there's a guy there from

Stetson-Ross Machine Co. and they're going to start that planer up and there's nobody down there. So I went down. What a nightmare getting it going. It was 75- and 80-hour weeks — a lot of time and money."

One of the more formidable headaches involved a three-story maze of pipes and cylinders that made up a "wood gasification plant." The intent was to burn sawmill and planer wood waste to produce gas, which would initially be used to heat the mill's dry kilns and would later supply all the plant's electrical needs.

David had ambitious plans for the gasification plant. Once the design was perfected, he envisioned selling similar units to other mills for half a million dollars each. A separate company, Westwood Polygas Ltd., was even created to handle the sales and manufacturing activity.

The concept for the plant had been developed in 1973 by Moore Canada Ltd., which manufactured most of the natural-gas-fired kilns used to dry green lumber in B.C. After two years of research and development, Moore engineers figured their invention was ready for the market.

David Ainsworth, a self-described "sucker for something new," bought the first unit for $480,000 on assurances from Moore that they'd have it running within a year. The technology would be a huge benefit to the Chasm mill, which was too far from the Westcoast Transmission pipeline to bring in natural gas.

Before the year was over, however, Moore had to bail out of the gasification project when its U.S. parent company ran into financial difficulties. Moore bought out of its agreement with Ainsworth for $100,000 and turned over all the rights and patents. Ainsworth also received the services of two Moore engineers.

"At the time I was still enthusiastic about it because we now had some experience

David Ainsworth had high expectations for the gasification plant. He hoped to manufacture and sell units to other industrial users.

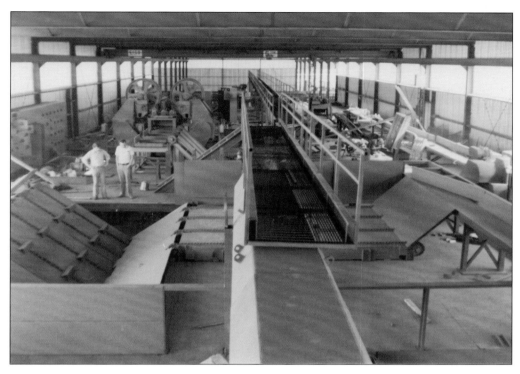
The new Chasm sawmill was designed to produce nearly 50 million board feet of lumber a year.

running it, and we had had some measure of success," says David. "The intention was to run the plant continuously without a lot of people hanging all over it."

Burning wood fibre to make producer gas was not a new concept, but it had never been attempted on this scale — and on a commercial basis at a sawmill. Wood waste with chips and sawdust was difficult to handle at the best of times, and figuring out how to feed the correct measure through a pressurized conveying system and into the plant's firebox was only one of the hurdles that had to be overcome.

Getting the gasification plant to run properly was further complicated by the involvement of several federal government ministries that had partnered on the deal.

"It took a lot of money to build it, and everyone who put money in thought they should run it," says Dick Sellars, who eventually became Chasm's first site manager. "The biggest problem was caused by the federal government sending truckloads of people there. They'd say, 'Let's see if we can burn garbage in it.' So they'd fill the thing with garbage and screw it all up. Then that wouldn't work so they'd try something else. The mill turned out to be a place these guys figured they could go and practise with this technology."

At peak performance, the gasification plant ran for 70 consecutive days on a couple of occasions, but it was far short of what the mill required.

"The plant had tremendous potential," says David, "but we started to run out of money, and I was gullible enough to sign the agreement that put these federal guys in charge. I thought they'd be useful once we got the thing going and we were selling units all over the country. In hindsight it was a very foolish thing to do. None of us could put in enough time on the thing because we were in the sawmill business. This was taking a lot of time and money and effort."

By the end of 1977, Chasm had been converted from a dimension lumber producer to a stud mill.

Much to David's disappointment, work on the project was finally halted in 1981. Within several years the structure was torn down and sold for scrap.

In the meantime, Chasm's kilns had been heated with bunker fuel, which was inordinately expensive. With no prospect of heat energy coming from the gasification plant, the Ainsworths salvaged a hot water boiler system and additional kilns from an operation at Northwood Upper Fraser near Prince George and converted Chasm's existing kilns to the hot water system as well.

Problems in the mill's first few years of operation weren't limited to the gasification plant.

Sawmill waste that didn't make it into the gasification plant initially had to be trucked off site to a landfill. A beehive or teepee burner from Prince Rupert was dismantled and moved to Chasm. Although it was a welcome addition to the operation, it required an extension almost immediately to prevent copious amounts of fly ash from escaping. The burner eventually had to be replaced when production grew to 100 million board feet annually, producing more waste than the burner was designed for.

Building the sawmill and planer on a single foundation also proved to be a mistake. Crews soon discovered that vibrations from the chipper affected the planer's performance. The only solution was to make the planer's foundation independent, a difficult task to carry out in tight quarters while keeping the mill operational.

Meanwhile, plans for pressure-treated lumber had to be abandoned soon after startup; it seemed that the hard Interior lodgepole pine didn't allow for proper penetration of the chromate copper arsenate solution. The plant did see some use with the eventual production

> One of the carloads of the company's lumber completed a round trip to Alabama, U.S.A. The car was loaded and left 100 Mile on October 6, 1975, and went all the way to Alabama as ordered. However, because of some oversight, the car was returned loaded to us at 100 Mile House. W.E. Hunt and Leo Poirier did a good job of loading as the car arrived back in 100 Mile after its 6,000-mile voyage in perfect condition.
>
> — THE AINSWORTH NEWS, NOVEMBER 1975

Chasm also produced pressure-treated posts.

of treated posts and poles. Although never big money-makers, posts reached a production high of 150,000 units annually before being phased out in the early 1980s.

Chasm's most serious drawback, however, involved the log supply. The mill was built as a random-length sawmill based on initial timber assessments that put the average log length at 16 feet. In reality, the average length was closer to 11 feet. The wood coming into the yard was generally skinnier and more crooked than originally anticipated. The result was more waste and poorer recovery.

At the same time, the market for random-length lumber was in a downturn, especially the precision-trimmed products Chasm was able to produce for the mobile home industry.

By the fall of 1977, barely 18 months after startup, the Ainsworths decided to convert the mill to a stud mill. By sticking to eight-foot lengths, there would be less wastage. Also, studs at the time were selling for higher prices.

The mill conversion took place over a frigid Christmas shutdown.

"It was quite a feat to go out and say we're going to make studs after Christmas," says Al Smith. "We had to chop stuff up and weld stuff in. We tried to do as much as possible ahead of time. In the meantime, we had the Exeter mill half torn apart, changing the set works on the Chip-N-Saw so that we could run railroad ties through there. We had a bunch of the crew up there working over Christmas because we had to get that mill going. It was our main mill, our bread and butter, and it had to go.

"Dave came in to the Exeter mill at Christmas — it was 25 below on a Saturday night and he wanted to know what we were doing. He thought we were crazy. We told him what we were doing and he said, 'You can't run it when it's this cold anyway, so don't worry about it and go home.'

"Well, we didn't go home. We finished it. Then we went to Chasm the next day and it turned 40 or 45 below. There was no heat in either of those mills. The guys were trying to unroll their cutting torch lines off the truck and they would just snap the hose, it was so cold. But we were able to get Chasm going just after New Year's."

Converting the Chasm division to a stud mill entailed major changes to the log decks and belt conveyors, but the sawmill equipment could remain intact with some alterations to the set works.

At the Exeter saw-filing shop. From left: Wilbur Cornthwaite, Tom Ainsworth, Norm Wood, John Code.

Chasm's twin band saw, manned by Leo Buis.

The reconfigured mill had several key attributes at the time. For one, it was able to produce 2×6 studs, which were now being mandated for new home construction, especially in the northern U.S. Roughly 90 per cent of the mill's output was divided between 2×4 and 2×6 studs.

The mill also had the ability to precision end-trim to any length up to 96 inches. Most lumber producers were limited to the standard-length stud of 92⅝ inches. Chasm, however, could precision trim for markets from North Carolina to California to overseas, all of which preferred varying lengths.

Logging roads in the Chasm area east of Highway 97 were virtually non-existent, which meant Ainsworth had to embark on an ambitious road construction plan. After all, some 40 per cent of their wood for the mill would come from this area. Larry Duncan, who had joined the company's road maintenance crew in 1968, was now the southern division's right-of-way foreman. His six-man crew would log the proposed route before arrival of the construction crew that actually built the roads and landing sites with a collection of Cats, loaders, gravel trucks and graders.

In less than two years, Duncan's crews put in over 110 miles of roads and nearly 450 landings in the Chasm operating area. Moving farther to the southeast, the road building continued with another 35-mile stretch in the Hihium Lake area. Major off-highway thoroughfares like this were called "systems roads." They were expensive undertakings, but forest licensees would eventually recoup the construction costs through stumpage offsets.

One of the more challenging projects in those days was the construction of a log bridge that spanned the Bonaparte River. Designed by Doug White and constructed by Robin Nadin's right-of-way crew, the 50-foot bridge was built to withstand loads of 150,000 pounds.

Like its sister mill in 100 Mile, Chasm also produced wood chips for the Interior pulp market, as required by the provincial government back in 1965. But by the early to mid-1970s the price for chips had slumped to the point where it wasn't even worth shipping them. The result was a huge surplus, with mountainous piles growing steadily at sawmills throughout the Interior. Operators had little choice but to stockpile and then burn the chips. Ainsworth, which shipped roughly 10 cars of chips a day, had amassed a pile that would have filled 300 railcars.

Prices were still being set by the pulp mills themselves, creating an untenable situation for the sawmills that supplied the chips. At rates of less than $10 a bone-dry unit, pulp mills were paying far less than what it would have cost them to produce their own chips.

When chip producers pleaded their case to the provincial government, export rules were eventually relaxed so they could sell their product to offshore buyers. Small quantities of chips from one producer were a tough sell on the international market, however, so Ainsworth joined with other Interior mills to form a chip export consortium called Fibreco Exports Inc.

> When I first joined the Mountain Pine company in the late 1970s and had the opportunity to drive up to visit our mills, I would pass through the little town of Spences Bridge along the way. After Spences Bridge you start to climb up onto the Cariboo plateau and there's a number of long hills. Well, at some point there was a truck delivering barrels of our turquoise paint, and one of the barrels had developed a leak. The result was that the paint dribbled right out of the truck, so down the middle of the right-hand lane going uphill was a turquoise stripe for some 23 miles.
>
> That paint smear was there for a number of years. We joked that if you ever wanted to go to Ainsworth you just follow the turquoise smear on the highway. David used to say it burned him up whenever he saw it because the company probably got charged for the paint anyway.
>
> — ROD LEE

Ainsworth "A" (far left) was used in the mid-1970s. Ainsworth's "Flying A" logo was introduced in 1978.

Ainsworth acquired the McMillan brothers' mill at nearby Lone Butte in 1978.

Within two years of incorporation, Fibreco had constructed one of the largest and most modern chip-handling facilities in the world. Located on the north shore of the Burrard Inlet, the depot was soon able to handle two million tonnes of chips annually.

Ainsworth quickly became one of the principal shareholders, based on the amount of chips the company shipped to the coast by railcar. Chip-producing mills farther north incurred higher transportation charges, while Ainsworth's relatively southerly location guaranteed its shipping costs were among the lowest of the Interior chip producers.

With Chasm now converted to a stud mill and Exeter's annual output increasing every year, Ainsworth took advantage of favourable market conditions in early 1978 and branched out with the purchase of McMillan Contractors Ltd., a sawmill located near Lone Butte, about 25 miles southeast of 100 Mile.

The mill was run by brothers Jim and Glenn McMillan, whose family had a long history in the area's forest industry. With approximately 70 employees, the mill had a reputation as an above-average producer with good-quality equipment.

Even more important than the mill itself was McMillan's timber quota, which was part of the $2-million selling price.

The regulatory system that favoured consolidation in B.C.'s forest industry hadn't changed significantly over the years, which meant that Ainsworth was able to acquire the cutting rights to approximately 40,000 cubic feet of timber annually.

> *I would periodically pop in at Exeter and visit Dave and I would sit down and talk, and maybe one of the boys would show up, and we'd start to talk. Then all of a sudden Susan would be on the scene, and you could see by the look in her eye, she would sort of raise her eyebrows as if to say, "Haven't you guys BS'd enough? Conrad, why don't you go look at the plant, David's got this and that to do." She would order us around pretty good and we'd respond, believe me.*
>
> — CONRAD PINETTE

The timber quota was based in an area east of 100 Mile House near Bridge Lake and Bowers Lake, featuring higher-quality stands of pine, spruce and fir than the company had been accustomed to. The acquisition opened up a whole new area of logging operation for Ainsworth. With Lone Butte added to the company payroll, Ainsworth had grown to over 350 employees, in addition to several dozen area logging and hauling contractors.

There were larger operators in the Interior, and certainly those with deeper pockets, but Ainsworth Lumber had overcome some significant obstacles to carve out a niche during its first 25 years in business.

David was now in his mid-fifties, with an ever-widening range of responsibilities — a directorship with Fibreco and participation in various industry organizations — but he continued to keep as close a contact with the mill and bush operations as time would allow.

His role as a community leader had been recognized several years earlier, in 1973, when he received 100 Mile's "Good Citizen" award. The local newspaper cited his participation in a long list of activities, from search and rescue to establishing the community's first hospital, serving on the pre-incorporation village commission and later as alderman for several terms, helping to found the local ambulance society, and a host of other contributions in time and money.

> *The new office is looking good with the windows in and the siding on. The wiring is roughed in and the drywall will be next. Lynne Plewes or Bill Becker will make no comment on when it will be finished, and Ken Greenall just grins.*
>
> — THE AINSWORTH NEWS, JANUARY 1978

Back at the Exeter office, Susan, who had been assisted for a number of years by Brian's wife, Eiko, and Allen's wife, Hiroko, now oversaw a staff of seven in a spacious new building constructed in 1978 at the Exeter site. Catherine Ainsworth was recently out of college and had begun her career at the company, working in the office as well.

Allen divided his time between overseeing mill operations and business development, as well as being directly involved in sales and marketing, while Brian continued to focus on woodlands activities.

As the family members knew only too well, the road thus far had been filled with bumps and potholes, all within a cutthroat industry that was rough-and-tumble at the best of times. And yet, despite everything they had been through and all they had accomplished, Ainsworth Lumber was only getting started.

Chapter 6

READY FOR TAKE-OFF

Skis in the winter and floats in the summer provided access to the Cariboo's abundance of lakes with the Cessna 185.

Whether he was piloting a single-engine Cessna, a Jet Ranger helicopter or a Citation Bravo jet, David Ainsworth loved to fly. His countless hours spent in the cockpit over nearly four decades in the air are tightly woven into the growth and development of Ainsworth Lumber.

If David wasn't scouting out timber stands or ferrying woodlands crews to remote corners of the region, he was at the controls on dozens of search and rescue missions, medical evacuations and mercy flights — many of which were on company time.

Being able to leave the pickup trucks behind and survey the region's forests from the air proved to be invaluable when bidding on timber sales. Aside from Pete Nadin and Doug White, who often accompanied David on these aerial surveys, it would be difficult to find anyone more familiar with the area's topography.

Getting your pilot's licence back in the 1950s and 1960s in the Cariboo was almost routine. In fact, local authorities claimed there were more planes and pilots per capita in 100 Mile House than anywhere else in Canada. Rudimentary airstrips could be found just about anywhere there was a flat piece of ground, from farms and ranches to small communities up and down the line.

The company's Cessna 185 heading out from 100 Mile's airstrip with a forestry crew.

David was nearly 40 years old when he got his pilot's licence in 1960.

"At the very beginning, I had always wanted to go flying but we didn't have the extra money," he says. "Then one day we were at our little office up at the plant and the phone rings, and it's Doug Theno wanting to talk to me. He said he was going to take flying lessons at the old aeroclub up in Williams Lake and asked if I wanted to go too. I said I'd be delighted but I didn't think I had the time or the money. I had to say no. So Susan asks, 'What was that all about?' and I told her, and she said, 'Well why don't you go?' I didn't see how I could spare the time, but she came right out and said I should go.

"So I did. Doug and I went up there, and after we got to the point where we were flying solo, we'd go up there at four or five in the morning to get in an extra couple hours of flying time before the other students came."

David was a quick learner in the cockpit and knew the test material cold, but that didn't seem to help when he took his written exam.

"There was a series of five questions along the lines of 'Which side do you pass another airplane on,' and the answer to each of them was 'right.' These were simple questions and I knew them, but I wanted to move on quickly to the tougher questions because I knew they'd take more time. So in my haste to answer those easy questions, I marked the wrong boxes. When the instructor and I found out that I failed the exam, it was unbelievable. You could have felled me with a matchstick."

It would be a long 30 days until he was permitted to take the test again, and this time it would require a trip to Vancouver. Much to David's relief, his second attempt was successful.

"So now I had a licence, and I wanted an airplane so bad I could taste it. Once I learned to fly, I started realizing this thing could be of some good if we used it in the right manner. It would be excellent for looking at timber or making the odd trip to Kamloops to see suppliers."

It wasn't long before David found himself at a dealership in Vancouver intent on looking at a used Cessna 172, the model he had seen others flying at the Williams Lake airport. The dealership's salesman must have recognized an easy mark; he marched right past the used 172s to a brand new Cessna 175. A short test flight later over the Fraser River and David was hooked.

The 100 Mile airstrip in the late 1950s.

"My heart was in my mouth before we had even taken off. I had trained in one of those old Fleet Canucks with a top speed of 80 miles an hour, so to me it was unbelievable what this airplane could do," he says. "It was like the difference between a Model T and a brand new Buick. And all the while I was saying to the guy that this was really kind of foolish because I couldn't afford this airplane. I couldn't really afford any airplane. I kept telling him this and when we landed he said, 'We'll see what we can figure out — maybe something on time.' First thing you know, I'm phoning Susan and telling her I'm going to be another week in Vancouver because they're teaching me how to fly this new airplane I just bought. I won't say I didn't get any flak from Susan on that one."

Allen and Brian started flying with their father almost immediately and before long were going solo themselves. Allen had already been flying for three or four years, since the age of 15, with his friend Dick Fish. By the time he finally got his licence in 1964 at the age of 23, he was an old hand at the controls.

Company aircraft made it possible to survey the timberlands throughout the region.

Several residents of Stettler, Alberta, were ferried by David in the 47-G to Halkirk's 60th anniversary celebrations in 1969. Over the years, more than a few senior citizens were treated to a helicopter ride courtesy of David.

The Cessna 175 still suited the Ainsworths' purposes when a representative from Bell helicopters showed up in 100 Mile in 1968 to demonstrate the new Bell 47-G3B2. David and the boys were definitely impressed after a few test rides, but there were some stumbling blocks.

"The Bell people proposed that we buy this thing, but it was a lot of money and we had plenty to do," says David. "We more or less chased them away, saying that somebody would have to teach us how to fly it right in 100 Mile. We might consider it, but we couldn't find the time to go to Vancouver to learn how to fly it.

"So away they went and in no time they were back, saying they had found us a man who would come to 100 Mile to teach us."

Since David had to leave shortly for a sales trip to Japan, he was first in line for instruction.

"It's always amusing when the instructor thinks you're ready to fly alone. With the helicopter we landed in a field out behind the Bridge Creek barn a whole bunch of times. I wasn't really expecting to go solo yet, and he says, 'Stop here for a minute.' So we land and he

David and grandson Douglas ready for a departure in the Bell Jet Ranger.

gets out for a minute — I'm sitting there in this thing and the rotor is going around and around — and he comes back with this big rock and puts it on the seat beside me. 'Okay,' he says, 'you're going to go solo now.' The rock was to make me feel comfortable, just to even the weight out a bit.

"So he says, 'Away you go,' and of course I went around a few times and finally I landed. He shook my hand and said, 'Now you've gone solo.' "

By 1972 the company had acquired a Cessna 185 with floats, which required another round of pilot certification. For an area like the Cariboo, which has an abundance of lakes, the float plane offered far greater access for inspecting stands of timber.

In the early 1980s, David and Allen advanced to Bell Jet Ranger helicopters. Not only did the five-seater Jet Ranger hold more people than the Bell 47-G; it also had far more horsepower and was capable of flying at 125 miles per hour.

The added mobility afforded by the planes and helicopters was a definite asset for the company. Air strips and helipads made it easy to visit the various divisions on short notice or cover a lot of territory in one day. Numerous trips were made over the years to visit other mills, customers or equipment manufacturers.

The community benefited as well. During the late 1960s and 1970s especially, local police and rescue officials relied heavily on the Ainsworths when aircraft were needed in emergency situations. The emergency response area around 100 Mile covers roughly 5,000 square miles, and the local ambulance service at the time was manned by volunteers, one of whom was David.

David in the Christen Husky, which was designed for aerial observation.

"There were commercial helicopter operators in those days, but they were very few," he says. "And the RCMP didn't even have a helicopter for some time. Later on, even if you could get a hold of the RCMP helicopter, they might say that they couldn't come up until tomorrow or the next day. So we would get called on lots of times because we were usually the only ones who could be in the air in minutes."

A month or two could pass without incident, followed by periods when an emergency call once or twice a week was not out of the ordinary. Aside from the occasional payment from the RCMP — their budget for such activity was virtually nil — the vast majority of the work was done for no charge.

Many of the calls were to remote areas in the bush or near a lake. Poor weather conditions often made the flying hazardous, especially in winter when thin ice or wet snow could make for dangerous landings on frozen lakes.

In other cases, those on the ground would have to start a fire to signal their whereabouts in the middle of dense forest.

Attending at logging accidents, boating mishaps and a host of other emergencies and transport situations was all part of the activity. Helicopter rescues over the years involved a stranded operator on the Big Bar ferry, stranded anglers on the Mahood River, and a stranded pilot and his passengers on Quesnel Lake who picked a stormy Christmas Day to require a lift out.

On occasion, there was even a bit of crime fighting involved, like the time David flew a couple of undercover RCMP officers to a remote site near Dog Creek where they collared a man for cattle rustling.

Another memorable trip included a highly agitated police dog on its first helicopter ride. The canine was so keen to get out of the chopper that it nearly jumped through the door when they started to land.

Various rescues involved people with life-threatening injuries who had to be wrapped in blankets and then strapped into a first-aid basket outside the helicopter. In later years the larger Jet Rangers made it possible to accommodate stretchers inside the aircraft.

If a rescue involved injuries that proved too severe for treatment at the 100 Mile hospital — a severely burned woman at Eagan Lake was one such case — David would ferry the victim to the hospital in Kamloops, an additional hour's flying time.

In the more unfortunate cases, there were fatalities to transport.

"Most of the people came out alive," says David, "but a few of them didn't. Some were pretty gruesome. It's never very pleasant when the person you're trying to save doesn't make it, despite everyone's best efforts."

Former RCMP member and ambulance chief Scotty Ramsay accompanied David on numerous emergency flights over the years.

"We'd phone Dave wherever. If he was in town he wasn't hard to track down. We'd call up to the office and tell them what the situation was and they'd find him for us. You never got any stonewalling. When someone was in trouble there was never any question of who's going to pay for this, who's involved. It was more like, 'Where are you and how am I going to find you?'"

By the mid-1980s, after hundreds of hours spent piloting single-engine planes and helicopters, David was ready to take his flying up a notch.

A big notch. The company bought a Cessna Citation I jet, and David was going to fly it. This was a huge leap for any pilot; the fact that David was 65 years of age made the achievement all the more remarkable. The number of pilots in Canada who have gone for their jet aircraft rating at such a late age can be counted on one hand. Commercial airlines like Air Canada require their pilots to retire at age 60.

Peter Shewring, who at the time was Chief Pilot for the B.C. government's air transportation service in Victoria, agreed to train David to prepare him for his test.

"At age 65, it's very unusual to go for that level of rating," says Shewring. "You just don't see it. It's tougher to learn the older you are."

David had extensive experience flying IFR (following Instrument Flight Rules — using navigational instruments rather than going by sight), not to mention piloting the company's turbocharged Cessna 210 Centurion. But he had never been licensed for a multi-engine aircraft.

The entire certification process took about a month, with roughly 10 hours of actual flight time in the Citation and 30 intensive hours of ground school.

"He would come down to Victoria when he and I were available," says Shewring. "He really wanted to fly that jet. We flew on the weekends, and he'd come to the Victoria airport where I could give him the classroom work. He was really determined to do it."

David's "good hands and feet" as a pilot, along with his mechanical aptitude, made a big difference in the transition to jet aircraft.

"I think he did exceptionally well in his training to get up to speed at that age," says Shewring. "He picked it up fast. And his knowledge of the mechanics of that plane was really impressive. If it was something with the hydraulics systems, for example, he knew it right off. He transitioned to the jet quite easily."

Taking possession of a company Citation jet at Cessna headquarters in Wichita, Kansas, in 1997. From left: mechanic Tom Schaff, pilot Peter Shewring, Cessna rep, David Ainsworth.

David gets ready to warm up the Citation on a wintry morning at the 108 Mile airport.

After David passed his official check ride from Transport Canada, he was certified to share Citation flying duties with a licensed co-pilot. Over the course of the next 13 years or so, the company's pilots included John Kite, John Downie, Peter Shewring, Howie Jamieson, Merv Hess and Brian Smedley, who also flew with Peter Shewring at Government Air.

The aircraft mechanic from the 108 airport, Tom Schaff, was also licensed to fly the Citation for maintenance purposes and spent many hours in the cockpit with David.

When the company twice upgraded to newer Citations in the 1990s, David passed additional levels of certification to keep pace with new electronic flight instrumentation. In addition to classroom work, these tests would include mandatory sessions in a flight simulator, which were often more challenging than flying the aircraft itself.

Having a jet aircraft certainly facilitated company travel, but those regular runs from the Cariboo's 108 Mile airport to Vancouver also helped out numerous Ainsworth employees with various medical needs.

Employees or family members with serious illnesses were often included on the flight list if it made a difference in getting to a hospital, medical appointment or treatment session, or helped them to visit a loved one. And as numerous guest passengers can attest, if their appointment went overtime, the plane's departure would be delayed until they returned to the airport.

David pared down his regular flying of the Citation with advancing years, and in 1999, his final year as an active pilot, he passed his check ride one last time with Transport Canada's Heinz Bold, who had tested him annually from the outset.

For someone who grew up behind a horse and buggy, it was one helluva ride.

David and co-pilot Brian Smedley in the Citation cockpit.

From the Cockpit: Looking Back

RESCUE AT THE BIG BAR FERRY

One Sunday afternoon Susan and I were about to take the helicopter to the Williams Lake airport to pick up Catherine, who was coming in on a plane. We got a panicked call on the phone from this guy at the highways department about some emergency at the Big Bar ferry. He wanted to know if we could go over there right away. It sounded interesting and exciting, so I said yes.

The Big Bar ferry goes back and forth across the Fraser River on a cable. It uses a rudder and is powered by the current. The ferry was shut down because it was flood season and the river was high. The current was very strong, probably going by at 30 miles an hour, and the main line was even dipping in the water.

The guy running this ferry wasn't supposed to be taking it out, but somebody had come along wanting to pick up a motorbike on the other side so they took the ferry over and picked up the motorbike. Then, about halfway across coming back, the ferry became unhooked and

At Susan's request, David once gave her an introductory lesson in flying a helicopter. After several hours in a wide open pasture attempting to hover, Susan decided that she'd leave the flying to David.

headed down the river. Still attached by a line, the ferry drifted close enough to shore that the motorbike guy was able to jump off, but the operator stayed on and drifted farther out where he got hung up on some rocks — and this river was really roaring along.

By the time we arrived, another helicopter from Lillooet was already there. The pilot was quite good and he had a plan. He flew out and dropped a line to attach to the ferry. Then he was going to pull the other end over to the west shore and tie it to a tree to keep the ferry from going anywhere, but it turned out the cable wasn't long enough, so this helicopter was hovering there on full power holding this line up, just short of the tree. It was hell to do.

I had some good rope with proper hooks on it, so we were going to tie my rope onto the cable so we could tie it off around the tree. But to hold onto the cable and tie it off, we needed more bodies.

Susan and I had come out with two highways employees, so I let Susan out and flew across the river to let them off. There wasn't really any room to land, so I kind of set down just below the other helicopter on one skid and told them when to slide across the seats and jump out.

We still needed more people. There were these three other guys just standing on the far bank with their mouths open, watching all this go on. I landed in front of them and motioned for them to get in, then let them out the same way on the other side.

And all the while this other helicopter was lying right over, just pouring on all the power he had to hold this cable. We finally got the rope tied onto the tree so he could let up his pull on it. By this time he was really getting worried about his fuel and was going to go back to Lillooet to fuel up.

I said I'd better try to get the operator off the ferry. This other pilot had more experience than I did, and he said, "If you go over there don't let the current fool you. Keep your eye on the ferry and don't look at the current because the river's running so fast it can confuse you."

I was willing to give it a try but he said that, on second thought, he'd better go himself and get the operator off. I had to haul all these bodies back and forth with no place to land.

Meanwhile, my poor wife was standing on the bank the whole time wondering what the hell I was trying to do, with all these guys hanging around. But she didn't complain.

The ferry was actually stuck there for a while. Then it got loose and went farther down the river. I don't know how they got it out of there, but they finally got it back in place.

— David Ainsworth

CLOSE TO THE TIPPING POINT

Someone had picked up a call on a portable radio that a police boat had gone down the Mahood River. Dave ended up taking these cops out to the river and one of them was huge — a really big guy. At one point they saw something and were about to land. It was a tight landing and the rotor was only clearing the trees by inches. You had listen to Dave if you were flying with him to get the signal of what to do. He'd always say leave your helmet and headphones on so you can hear him. Well this big guy just decided to jump out of the helicopter without giving Dave any warning. He was so big that the chopper just about tipped over and hit the trees because of the weight difference. Dave was very upset with him.

— Scotty Ramsay

WOODLANDS TRAINING FOR EVERY EMERGENCY

The Bell 47-G was pretty slow and you'd have to do switchbacks to climb over the hill going south out of town. We used to go out flying a bit with that, but the flying really kicked in with the Jet Ranger.

The bucket is filled in a nearby lake.

You almost needed to be certified just to open the door down at the hangar; there were only certain people who could touch it. If you pushed the button for too long then something would come unravelled.

David would often show up with new equipment like the "Bambi bucket," the big bucket that gets filled with water for fighting forest fires. This was in 1986. Dave would get everybody organized and we'd get half the woodlands staff together and we'd go out for a practice with the Bambi bucket, doing drills on how to hook it up to the helicopter.

This one practice was a classic: David was up in the air with the Bambi bucket and he pushed the wrong button. Instead of the water getting dumped out, the whole bucket just dropped to the ground. We all had to go running into the bush to find the Bambi bucket, and it was all smashed and wrecked. So then the Bambi bucket had to get fixed. I don't think we ever used the Bambi bucket on an actual fire, but we were ready if we had to.

— DOUG WHITE

Another load of water is dropped from the "Bambi bucket" in a 1986 training session for the woodlands crew.

ROUNDING UP A CATTLE RUSTLER

Old rancher Jack Alexander roared into the police station one day madder than hell and banged the tail of a cow on the counter. It was from one of his cows, he said, and some rustler must have stolen it.

An RCMP officer went out with him and they found the remains of two more animals. So two RCMP undercover officers, who handled all the cattle-rustling cases for the whole province, were brought in from Kamloops.

These guys were livestock experts, and the first thing they did was show up at the Twilight Lodge pub in a dirty green pickup with old cowboy clothes on. They were pretty skillful at talking people up, and pretty soon some people were telling them more than they probably intended.

Next thing you know, the RCMP knew who they were looking for and were able to go out and find the guy's place and even find the meat in his freezer. The guy worked at a portable sawmill out by the Sheep Creek Bridge on the other side of the Fraser River, but when they went out there to get him, he had left the mill, so they came all the way back to 100 Mile. Someone must have told them about me so they asked if I'd fly them out that evening.

They knew where this guy was getting picked up, on a shift change or something, and sure enough, there he was. They arrested him and we flew him back to town.

It was remarkable how they did all that in 24 hours when they started out with so little information. They were pretty smart guys.

— David Ainsworth

MUSTANGS MAKE AN IMPRESSIVE SIGHT

One time I was flying in the helicopter with David way out by Dog Creek, and in those days there was an actual herd of mustang horses there. Real, authentic wild horses, no brands or anything. We came over a hill and we came upon this herd, and this old stallion reared way up on his hind legs at us, just like it was out of a Walt Disney movie or something. Then he rounded up his harem and fired them off into the bush. That was really something to see.

— Pete Nadin

A MOOSE IN EVERY BUSH

A new customer, Al Norton, from Fort Worth, Texas, said he was coming up during Expo 86, so I introduced him to Allen Ainsworth. Allen flew him up to 100 Mile House in the helicopter, and they're going along and Al Norton asks, "Do you ever see any of those moose you've got up here?" Allen says, "Yeah, I bet you there's one in that bush right over there," and he flies over there and sure enough he scares a moose out of the bush. It was unbelievable. This guy is just in awe. He turned out to be a great customer, buying six-foot studs that were hard to get rid of.

— Blair Magnuson

The Vancouver Sun *of July 9, 1965, details the dramatic aircraft explosion near 100 Mile House.*

THE CRASH OF CP AIR FLIGHT 21

The 1965 crash of CP Air Flight 21 about 25 miles west of 100 Mile House was by far the worst disaster local residents had ever seen.

The DC-6B was en route from Vancouver to Prince George and on to Whitehorse in the late afternoon of July 8 when it exploded in mid-air and crashed into the bush, killing all 52 persons on board.

With their mill situated at the far edge of town, David, Allen and Brian Ainsworth, as well as many of their employees, were among the first people to arrive at the crash scene.

The Ainsworths and many others stayed through the night, doing their best to recover bodies and contain the scene while officials and trained personnel made their way to the site.

Investigators later determined the explosion was caused by a bomb, and several deceased passengers were pegged as suspects in connection to the incident.

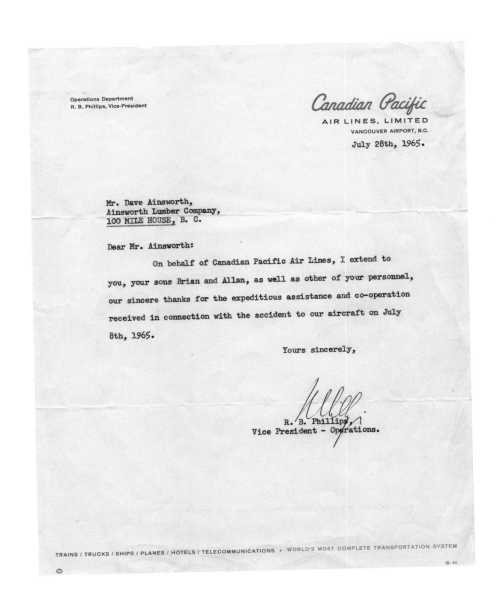

THE SEARCH FOR SAM KETCHAM

One of the searches I'll always remember happened when Sam Ketcham's helicopter went down in 1977. Sam had started up West Fraser Timber with his brothers Bill and Pete. I used to have a lot in common with him and got to know him quite well.

Sam, his timber man, Phil Bodman, and the pilot, Ken Waines, were in a hired helicopter up north. They were flying back to Quesnel. They were well along the way when they got caught in a snowstorm. Pretty soon they couldn't even see where they were going, and they didn't quite make it over a ridge. They all died in the crash.

By that time we had a helicopter, "The Bubble," on floats, and I took it up there over the next couple of days for the search. It was windy as hell and with those big floats underneath it was really leaning in the wind. So I phoned down to 100 Mile and said bring up the skids because flying this is getting to be pretty miserable.

Before I could get the skids on, though, word came that they had found the downed helicopter just north of Quesnel. By this time 22 helicopters had shown up for the search. It just gives you an idea of how well Sam was known, and how everybody tried to help. It was wonderful to see that kind of turnout, but it was a tragic end to the search.

— DAVID AINSWORTH

Above: Ainsworth's Bell 47-G3B2 had a cruising speed of 80 miles an hour. Right: The Jet Ranger was used in many emergency situations.

UP A CREEK WITHOUT A ROTOR

In 1970 our local Member of Parliament, Paul St. Pierre, asked me if I would fly him and Mitchell Sharp, Canada's Secretary of State for External Affairs, to this lake in the Chilcotin for a day of trout fishing. Mitchell Sharp was Deputy Prime Minister at the time and was in the area to open the Williams Lake Stampede.

So I agreed to take them out in the helicopter, the 47-G. On the way, we stopped at a ranch where we met up with the owner and two young guys who had driven there early that morning. One of them was an aide to Mitchell Sharp and the other was Conrad Pinette. They were going to go fishing too. We waited around for what seemed to be an endless time and the day was just wasting away. Finally I left for the lake with Paul St. Pierre and Mitchell Sharp. There was no clear spot to land right near the lake, but there was a little swampy clearing nearby. I didn't want to set down right in the middle of the mud, so even though I knew better, I tried to set down on two or three humps over near the edge, where it was drier. This was fairly early in my helicopter-pilot career and I miscalculated where this one tree was and ended up clipping it with the helicopter's rotor. It was a very small tree but it tore the plate off the end of one rotor blade. So now the blades were out of balance and we started to rock a bit. I got clear of the trees and set the thing down right away.

I checked the damage and knew I had lost some of the balancing weights along with the end plate. The skin on the rotor was damaged, broken open a few inches back on each end. This was going to take some fixing, and all we had with us was a fishing kit.

We found an old chair to stand on — of all the ridiculous things to find out there beside a lake — and tried to balance up the blades by adding something on. We even wired on some fishing-line leads. Before we could do that, we had to work some holes in there so we could wire it closed with fishing wire. We tried this kind of thing two or three times and the rotor still wobbled, so we finally took the end plate off the undamaged blade and hoped that would balance things up.

The mosquitoes were out there by the thousands; it was just horrendous. So there I was, out there with Mitchell Sharp, standing on a chair in the middle of a swamp, trying to fix a rotor above my head. It wasn't much fun.

I got Mitchell Sharp to pull some tape off this survival kit I had and we used that to kind of clamp down the ends of the rotor. We had to count how many wraps of tape we had on the end of each blade because they are so delicate. Even something like a paper clip on one blade would put them out of balance.

Every time we made a change I would have to start up the helicopter, test the rotor a little bit, then slow it down and do something else. It took quite a little while. The tape seemed to be holding okay, but the blades now made quite a significant whistling noise because of their open ends. It was a real "woo-woo" kind of sound.

Mitchell Sharp had a plane waiting for him back in Williams Lake, and I didn't know if the helicopter would make it back to town without some new piece flying off. So I flew by myself over to that ranch, thinking maybe a plane could fly in there and pick him up, but there was nowhere for a plane to land and it was starting to get late.

These two guys who had driven out to the ranch — the assistant and Conrad Pinette — were expecting me to come back a lot sooner and fly them out to the lake too, so they were wondering what had happened to me.

They had driven to the ranch in a pickup truck with a barrel of fuel and some of our tools, which happened to include some tape. So we took apart the blade ends again and retaped them and made them a little tidier. Then I refuelled and went back for Mitchell Sharp at the lake. I flew back over open clearings just in case something else flew off the rotor and I started wobbling pretty bad and needed to land.

When I got to the lake, Mitchell Sharp and Paul St. Pierre had managed to get in some fishing and actually caught two nice trout. We put them in a wet sack on the side of the helicopter.

We flew back to the ranch and then I was going to fly Mitchell Sharp back to Williams Lake. For some strange reason I agreed to take along the guy who owned the ranch, which was a big mistake. He was so full of himself he was talking the whole way, and I was trying to fly this thing with the taped-up blades and pay attention to what I was doing. Then he started telling me I didn't know where the hell I was going.

I didn't say anything and just kept on flying because I did know where I was. When the Fraser River came into sight, right where I knew it would be, this guy was just about screaming. He thought he knew the area, but his horse probably had a better idea than he did. We followed the river into town and I let him out at the Chilcotin Inn. I don't know what happened to him and I've never seen him since.

I let off Mitchell Sharp so he could get on his plane, and by this time it was almost dark. These taped-up blades were still *woo-wooing,* so I set off for 100 Mile and it was bloody dark by the time I got home.

Susan was out for a walk and was just going by the airport when I landed.

She said, "It sounded kind of funny when you came in, didn't it?"

I said, "Yeah, it sure did, and you don't know the half of it."

I was so embarrassed and ashamed of the damage I caused to that helicopter that I put it away hastily, and I didn't open the hangar door again for 10 days, until the new rotor arrived.

By the time we opened that door to get the helicopter out, those fish were getting pretty ripe.

— David Ainsworth

IN DEEP

Dave and I were out in the Bell 47-G looking for a plane wreck and someone had seen what they thought was wreckage. We were landing across this logging road in the wintertime. From a few feet up, the ruts in the road looked like they were made by a small car, but in reality something as big as a skidder had gone down the road. The snow was quite deep and these ruts were quite deep.

The helicopter had these big pontoons at the time which were about two- or three-feet high. So when I got out I had to slide down the pontoon first. Well, I kept sliding and sliding and sliding and there was still no ground below me, and I've still got my one boot stuck up on this rack above the pontoon. So finally I just let go and I fall straight down into this rut, which seemed about 10 feet deep.

The funny thing is I could actually hear Dave laughing over the *thwack thwack* of the rotors, which is pretty loud.

— Scotty Ramsay

HOUSE OF COMMONS
CANADA

OTTAWA

July 8, 1971

Mr. Dave Ainsworth
100 Mile House,
British Columbia

Dear Dave,

As a flying machine it's not bad, but it will never replace a power saw.

Mitchell Sharp had a great time and with the years this story will expand beyond the imagination of either of us.

Sincere thanks.

Best regards,

Paul St.Pierre, M.P.
Coast Chilcotin

Enc. (1 book)

The Secretary of State for External Affairs   Secrétaire d'État aux Affaires extérieures
Canada

Ottawa, K1A 0G2
July 7, 1971

Mr. Dave Ainsworth,
100-Mile House, B.C.

Dear Dave:

This is just a note to say "thanks" for the lift into the ranch and on to the lake for the fishing.

I was indeed sorry about the damage to the blade of your helicopter and I hope that by now you have been able to get it replaced.

If I ever decide to leave politics, I now have another field of endeavour open to me -- with a capital investment in one roll of scotch tape and some good fishing line, I can open up a helicopter blade repair shop!

It was a pleasure to have met you and I hope we will get together again one day.

With all good wishes,

Yours sincerely,

Mitchell Sharp

A FLIGHT TO REMEMBER

One night I was out practising in the 100 Mile Flying Club's Cessna 172. I was heading back to town and was over Horse Lake. Granddad was in his Christen Husky returning from Kamloops, where he had just had some new avionics and radio equipment installed. He chose a Loran C and King radios and avionics, which would allow him to fly in instrument-only conditions if necessary.

Both Granddad and Allen were big believers in the safety aspect of flying. Part of this meant having the training and the equipment necessary to fly in instrument-only conditions if the weather got bad, so I think he was pleased with the new equipment and eager to try it out.

Just before dusk, I heard him call his position on the radio as he approached the airport, so we started talking. Pretty soon we were flying in side-by-side formation down Horse Lake so that Granddad could check the groundspeed of his Loran C against my airspeed indicator. It was kind of an academic experience for him, but a great memory for me. I'll never forget flying side by side down the lake with my grandfather.

— KEVIN AINSWORTH

The paved 4,877-foot runway at the 108 Mile airport.

The Christen Husky's tandem seating also features dual controls with a stick rather than a yoke.

THE BEST KIND OF BACKSEAT DRIVER

During the summer of 1989 David had spent some time in the hospital, and he couldn't fly for a month or two. I had recently completed the training for my private pilot's licence, so we flew together. I got to sit in the front seat as the "chief pilot" and read all of the instruments while Granddad sat in the back. His new Christen Husky was a tandem two-seater, tail-dragging airplane modelled after the very popular Piper Super Cub. I got to do a lot of the flying, but I had a tough time with the landings because tail draggers are a little trickier for beginners to get on the runway.

A couple of times it seemed like I was bouncing all over the 108 Mile airstrip, which is where we went to practise. Granddad was in the back seat, patiently keeping us straightened out with the dual controls, which were the old-fashioned "stick" type. It must have been a bit disorienting flying from the back with my head in his line of vision, but he had no problem taking off, landing and correcting my mistakes from the back seat. We had a great time!

— Kevin Ainsworth

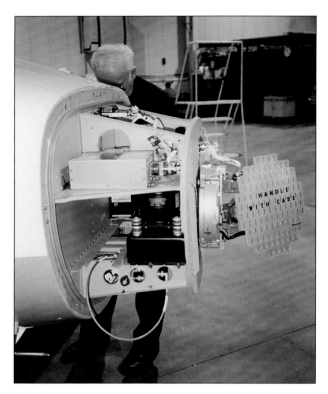
David pitching in on Citation maintenance.

GOING THE EXTRA MILE

Dave loved to fly. One day we flew up from Vancouver to the 108 in the Citation. Then we went out flying in his Christen Husky, and we ended the day by taking out the Jet Ranger helicopter.

He was strong, too. He'd come to the airport and run equipment, plough the snow. I'd work with him all day and he could outwork anyone. Before he got his ATV he would pull the helicopter out of the hangar by himself — and on snow. That's very hard to do.

Dave says if anybody tells you something is impossible, run away from them immediately. It was like that when we decided to do our own maintenance and inspections on the Citation right here at the 108 hangar. Everybody laughed at us and said it couldn't be done. I said we should do it here and Dave allowed me to do that. These Phase 5 inspections require 300 hours of work every two years or 1,200 hours of flying. We did them over Christmas when the plane was down. And you know what? We saved over $70,000 every time by not sending the plane out. Plus you'd really get to know the aircraft well.

I'd hire a crew and Dave would be right in there helping too. It would be 11 or 12 at night — and Dave is in his seventies by now — and he'd say, "Tom, I know you're going to stay a little longer but I'm sorry, I've got to go. I'm getting a bit tired." So here's Dave feeling bad about leaving at 11 after working with us all day.

— Tom Schaff

(108 airport manager Tom Schaff served as Ainsworth's aircraft maintenance mechanic starting in 1988 and carrying on into the 1990s.)

BURNED ON A DINNERTIME DROP-OFF

We were about 10 miles east of Bridge Lake, around Willow Lake, and there were six of us out there fighting a fire, using Cats and building a fireguard. We started about seven in the morning and it took all day and into the evening to get that fireguard way around the lake. It was getting dark and we were using lights to finish up. Dave knew we'd all be hungry and thirsty, so he flew out in his helicopter and dropped off a bunch of sandwiches and cases of pop so we could eat right there when we were done. There was a road that ran right along the area where we were working, and we could see him drop the stuff off right by Doug White's truck. When we all got there ready to eat, somebody had stolen everything. We were beat like you wouldn't believe, after 14 hours on the bloody Cat, and some guy driving by stole it all. I got home too tired to eat, went to bed and was up again at five to get my fuel ready for another day of work.

— JOE DANCZAK

Choppering out food to famished crews on the fire line was all part of a day's work.

RAPID DESCENT

One trip, we were flying home north from the U.S., and we were supposed to stop in Penticton to go through Customs. So we're flying along at 9,000 feet and I see Penticton down below and say, "Didn't we have to stop in Penticton?" Dave says, "Yup, but that's Oliver down there." So I said, "Whatever you think." Then Dave looks down again and says, "That is Penticton!" and banks the plane sharply. My ears popped pretty good — we went from 9,000 feet to nothing in pretty short order.

— DICK SELLARS

The company's first Citation aircraft.

SEARCH OF A DIFFERENT KIND

We went to visit Sunrise Lumber in Spokane by flying down in the Cessna 210. The customer picked us up and took us to the plant. Everything was good; they were buying a lot of 1×4 and were happy with the quality.

On the way back one of the guys was interested in this airplane, so Dave said, "Come on, I'll show it to you." He opened the door and the door swung right around to the engine compartment. Everyone was crawling around, looking at the aircraft, and the guy was really impressed. Then he drove away and left us there.

Now, Dave always has a million sets of keys for everything, but this time he couldn't find the keys to the airplane. Not in the dash, not in the vest pouch, not in this coat, not in this briefcase or that briefcase, not under the seat … so we were there for quite a little while looking for these keys. Finally I said, "How did you get the door open if you didn't have the keys?" Dave peeled the door around and there were the keys, in the door.

— Dick Sellars

AN UNPLANNED RESCUE

David and I were in the helicopter, flying over all of our southern timber supply. We flew out west and east, down to Savona, back toward Cache Creek, and on the way back, near the Hat Creek Valley, we were way back in against some steep mountains, circling around and starting to head north when I saw some motion down below. David circled again and we saw something white.

This was a very tight canyon with tough conditions, and there was a car at a landing. Near it was a woman. David set down, we got out, and the woman told us she had been there for two nights, sleeping in the car. She had been given some bad directions on the logging road and went into the landing. I guess the landing punched out and the car sunk down.

She thought we were part of a rescue party, and we told her that we had found her by accident. It would have been a long walk out for her.

That was one of David's rescue missions that wasn't planned.

— Steve Silveira

MERCY FLIGHT TO ONTARIO

When Ainsworth Chasm employee Willie Roose, 43, was told he had bone cancer late in 1986, his first concern was for his two young daughters, Vicky and Kimberly. His condition was rapidly deteriorating and, as a single father, he wanted his daughters to live with his brother and his family in southern Ontario.

By this time, hospitalized in Kamloops, his last wish was to spend his final days in Ontario with his mother, brother and daughters at his side. Unfortunately, the family was unable to afford the commercial airfare since Willie alone required three seats for his stretcher.

Time was quickly running out when David Ainsworth was contacted about the situation, and he didn't hesitate to offer his services. The following day he and co-pilot John Downie flew Willie Roose, his mother and his two daughters to London, Ontario, on the company jet.

By that night, Willie Roose was in the Tillsonburg hospital and his daughters were with his brother Leo and his family.

"It was a very sad situation," says David. "We decided to take him over as soon as we were contacted. Willie was a loyal employee of 10 years; it was the least we could do."

Willie Roose's mother and daughters were passengers on a flight from B.C. to London, Ontario.

SIX-DAY SEARCH FOR LOST BOY

The most important search I was on — certainly the most exciting and best organized — was for a 12-year-old boy called Jeremy Morton. He was Millie Hamilton's great-nephew, and he went missing out around Murphy Lake, northeast of Lac La Hache. It was July, 1983 and he was out for six nights by himself. The kid had been out there two nights by the time the RCMP asked us to come in with our helicopter.

I had Doug White and Robin Nadin with me and we were out there until the end. It seemed like there were hundreds of people involved. If a rancher had a pickup truck with a rack around it, he'd put a horse in there and away they'd go. People were out searching on horseback and on foot.

We'd go back and forth and back and forth. The first day out there we found the boy's pony, so we thought we were on the right track. I guess the pony had got away on him. We figured we had to find him soon or he'd be in pretty rough shape. But that kid kept on moving. He kept heading east to the end of a road where they had a logging camp.

He was found flopped out on a bed when the caretaker couple came back. I guess he was just exhausted. The RCMP got the message first, so they went in there to get him. We flew in the mother and grandmother soon after so they could be right at the scene.

We were so intrigued with that search that we just didn't want to give up. And the people from the Lac La Hache area were so supportive. They'd send in a pickup truck filled with boxes of sandwiches and everything for all these people out there at the search site. There was more community involvement with that search than anything I've ever been involved in. It was the most gratifying search I was ever on.

— David Ainsworth

Above, left: A brief time out for pilot David Ainsworth and Doug White during the search for the lost boy. Doug recalls, "I'd come home from that search each night completely exhausted. It's hard work flying around in that helicopter all day. I also had the map and was supposed to know where we were all the time. Dave would be ready to go at six the next morning. He had tremendous stamina, and he wasn't a young man at this time. That was quite a week." Above, right: The base camp near 2 Mile Lake attracted hundreds of searchers and volunteers.

READY FOR TAKE-OFF

Chapter 7

MAKING THE GRADE

B.C.'s new Forest Act in 1978 increased the emphasis on the stewardship of the forest resource.

Considering the Ainsworths' "no brakes, no reverse" approach to running their lumber business, no one could accuse them of growing comfortable with the status quo. In any event, developments in B.C.'s turbulent forest industry during the 1970s didn't allow for a lot of "standing around with your mouth open," as David would put it. If the changes weren't occurring in the bush or at the sawmill, they were happening in the halls of the legislature.

In 1978 the Social Credit government brought in a new Forest Act in response to a Royal Commission report completed in 1976. The report itself resulted from a massive chip surplus in 1975 that made the NDP government realize provincial forest policies were in dire need of review.

Undertaken by University of British Columbia forestry professor Peter Pearse, the one-man Royal Commission covered a wide range of issues, including his concerns over tenure policy and the concentration of cutting permits into fewer and fewer industry hands. The report also heralded a shift in thinking about B.C.'s forest resources. Rather than focusing on the continuation of timber production, Pearse recommended that new policies should put more emphasis on conserving and enhancing the forests. Changes were being sought from within and outside the industry, and Pearse's report had successfully captured a wide range of opinions.

In releasing its new Forest Act, however, the Social Credit government chose to ignore some of Pearse's key findings, not the least of which involved the corporate concentration of cutting rights. A new type of tenure based on Forest Licences was introduced, which only made the concentration of cutting rights more inevitable. Not surprisingly, the new Act's formulators didn't bother to consult with Ray Williston, the forward-thinking former Minister of Forests.

The new Act did take note of Pearse's recommendations on forest stewardship and the recognition of changing attitudes around the world regarding environmental issues. As a result, the Forest Service would now take into consideration the care of wildlife, fish, water and recreation potential when it made a timber-related decision. In addition, the corresponding ministries would have input as well.

Putting the new Act into practice did not go smoothly. The bureaucrats in charge decided to use the occasion as an opportunity to reorganize the Forest Service. The result was a much more centralized ministry; in many cases the seasoned forest rangers were reassigned, while senior positions went to those with academic credentials.

What used to be routine decisions now required input from specialists in a variety of fields rather than from a single forester with more generalized knowledge. Within several years the system was in a shambles and the massive reorganization was halted.

With or without the new Act, getting decent timber to Ainsworth's mills continued to be a challenge. In some cases, cutting rights were spread over land controlled by numerous types of permits with varying stipulations for the timing of the cut, as well as for the quantity and quality of the wood to be harvested.

Ainsworth had originally described the timber quality in the Chasm bid as "below average," with 90 per cent of the pine measuring less than eight inches in diameter.

> *It became really apparent early on that we had to stick together. On so many issues it was like the Ainsworths against the rest of the industry — if you just turned your back on some of the competition, they'd eat your lunch. So it was a struggle just holding our own. But if you held your own long enough, you moved forward.*
>
> — ALLEN AINSWORTH

"Some of our licences were non-renewable over a one-year period while others were over a five-year period," says Pete Nadin. "They were a use-it-or-lose-it type of thing, so it became quite a balancing act sorting out where and when to log. You only have so many resources to get this all done. At one time we had 13 timber quotas that were operational."

The late 1970s and early 1980s also saw a massive infestation of the mountain pine beetle in dry-belt regions west of Highway 97. The area's relatively untouched stands of mature lodgepole pine were prime targets for the beetles. Extreme cold in the winter or forest fires in the summer would normally have kept the beetle populations in check. But winters had been getting progressively warmer, and forest fires were rarely allowed to spread. Conditions were ripe for the beetle's attack.

Affected forest licensees such as Ainsworth had no choice but to log vast swaths of beetle-infested timber in an effort to control the outbreak. Although spread over a number of years, much of this cutting was done at the expense of more valuable stands of timber that offered a better profile of wood for the sawmills.

"We had to focus our harvesting in the areas that were affected," says Pete Nadin. "The wood profile where the infestations were taking place wasn't necessarily the best, so we were getting a lower quality of log."

Vast swaths of lodgepole pine in the dry-belt region west of 100 Mile House were destroyed by the mountain pine beetle.

Chasm's conversion to a stud mill, meanwhile, went a long way toward boosting wood recovery and reducing waste, but there was always room for improvement. It just so happened that the next step in the process stretched all the way south to Texas.

Allen Ainsworth was in the southern U.S. on a sales trip when he was introduced to "finger-joined" lumber. The product is composed of trim blocks or short lengths of useable lumber that are machined with a finger profile at either end, then glued together to form a continuous board that is cut to the desired lengths. The finger-joined lumber sizes included 2×3, 2×4 and 2×6 boards with varying lengths up to 12 feet. The trim blocks average around 18 inches but can be as long as 36 inches.

Not only is potential waste material converted into a value-added product, but the resultant lumber is more stable than natural studs. This increased stability was particularly important in markets like Texas, where extreme temperatures can warp any lumber that has excess moisture content.

Ainsworth's office at the Exeter site was a hub of activity.

The finger-joining mill equipment from Texas arrives at the Exeter site.

With Ainsworth's ready supply of finger-joining stock from Chasm and Exeter, it made good business sense to start up a finger-joining operation. The necessary equipment was purchased from a manufacturer in Texas and shipped to 100 Mile House. Pinette and Therrien, one of Ainsworth's partners in the Mountain Pine Ltd. sales co-op, put in a similar mill around the same time.

Attempting to manufacture finger-joined lumber in a colder Canadian climate was a bit of an experiment and not without difficulties.

"It was a battle from the get-go," says Dick Sellars, who was now overseeing quality control at Chasm and Exeter.

One of the first hurdles to overcome before production could even begin was obtaining the necessary grade stamps and product certifications. Involving a battery of tests and trials at the Western Forest Products Lab in Vancouver, the work began at least six months before the 1978 startup and carried on for several years after product certification had been gained in all the key market areas.

"There were no finger-joiners in the country yet, so who's going to produce a grade stamp?" says Sellars. "We had to go down and argue with the scientists at the lab about producing this standard, and they weren't very willing at first. There were always more tests they had to do. We were in a meeting there and I said, 'Next Monday this plant is starting up and I'm going to have to put something on those boards even if I have to make it myself.' And I walked out, slammed the door, went home, and the following Friday they phoned and said we could use the natural stud stamp and put 'Finger-joined — Vertical Use Only' on it."

The mill itself was constructed on the west side of the planer mill at Exeter. Al Smith, who had never seen a finger-joining mill in operation before, was responsible for setting it up.

It didn't take long to discover that producing finger-joined lumber in the Cariboo came with its own set of challenges once winter rolled around.

The mill was unheated, which meant the glue didn't set up properly in the finger joints when temperatures turned colder. Boards with defective joints could simply fall apart, and if it happened to a customer on the job site, there was hell to pay back in 100 Mile. The problem was partially solved with the addition of a large propane oven to warm the pieces before gluing, but this added to the production costs.

"That was the fun part about working for Ainsworth, though," says Dick Sellars. "There was a fair bit of autonomy, and we were encouraged to come up with solutions when we hit a stumbling block. And so you'd have guys like Don McGladdery, and he'd say, 'You know the

The finger-joining mill was constructed on the west side of the existing planer mill.

wood just won't set, so we should try to warm it up somehow,' and we'd say, 'Well let's build an oven — anyone know how to do that? Oh we'll get someone at Exco,' and they'd make an oven, and that worked better. That's how the thing evolved into a pretty damn good plant."

Spare parts, meanwhile, often had to be shipped from Texas or were difficult to locate. In many cases it made more sense for Ainsworth shop employees to fabricate their own.

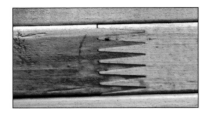

Don McGladdery, who with his wife Sue joined Ainsworth in 1979, oversaw a crew of 20 employees on the finger-joiner and was as familiar as anyone with the quirks of the machinery.

"After the pieces were glued together, they came out in a continuous board where a saw would cut them into individual boards," says McGladdery. "We had a limit switch at the saw but sometimes it would miss a cut, so then we'd have a board going out that was 200 feet long before someone noticed it, because there was no operator in that area. It happened maybe once a week, so we actually cut a hole in the wall with hinges on a cover. That way, if the switch didn't go on and we produced one of these 200-foot boards, it would go right out the wall and into the yard."

If excess glue happened to squeeze out of the joints during packaging, an entire tier of studs in a bundle might be stuck together. Then the customer would be forced to break the boards apart with an axe.

Early production problems notwithstanding, the finger-joined lumber was soon drawing a premium of $25 to $50 over and above the usual return for a thousand board feet of natural studs — and all this from raw materials that would otherwise be sold as pulp chips for a much lower return.

Long-time employee Eugenio Cargnelutti performs a strength test on the finger-joined lumber.

By the early 1980s, production at the Exeter stud mill had increased significantly with three kilns in operation.

"Texas was pretty much the first market we shipped to," says Dick Sellars, "because they hadn't really heard of finger-joined lumber in any other states. California was the second state to come on line."

Despite the addition of the oven and other innovations, cold temperatures during winter made production at Exeter tenuous for up to five months of the year. By early 1986, plans were underway to move the finger-joiner to a site in the Lower Mainland, with its milder climate.

In the meantime, production at the Exeter stud mill had seen a steady increase, and by 1980 nearly 80 million board feet were being produced annually, up 60 per cent from startup days back in 1970.

The endless flow of logs and lumber through the mill was constantly being scrutinized in an effort to extract the most value from the resource.

"We were able to pump up the throughput at Exeter, and later Chasm, just through rebuilding things and making improvements on what we had," says Al Smith, who was now operations manager for all three of Ainsworth's sawmills.

"The ideas came from all over the company, from the crews and from Exco people building the equipment. There was a lot of working together; you'd put something in, try it, and develop new ideas if you had to, like ways to improve the log handling or the log deck or the Chip-N-Saws themselves. All those things made a difference."

The Exeter mill also began producing railroad ties with the addition of a scragg saw, a tie stacker and a tie trimmer.

"There were times when you couldn't sell other products but you could sell rail ties," says Smith. "You had to have the logs that would go with them, but you could make a good dollar at those."

Long-time employee Jim Keller (on fence) heading up a work crew.

Accepting an award from Seaboard. From left: Jerry Daykin, Bob Glover, David Ainsworth, Dick Sellars, Bruce Adams.

By now the Ainsworth studs, with their turquoise end-paint, had gained a reputation in markets around North America as a premium-quality product.

Long-time lumber salesman Rod Lee noticed the difference when he first joined Mountain Pine Lumber to sell Ainsworth's product.

"When I took my very first trips up to the mill, I was introduced to a lot of the mill management people, the mill quality-control people, and in every instance these people had been trained by David to put out a product that was better quality than the minimum standards," says Lee.

"It was a very deliberate move on David's part. If the market wasn't good, and if we ever had to come down in price to compete with other products, David felt we would always get the order first because of our quality. That was the way to maintain cash flow even when the markets were down. And it worked. Customers were always prepared to buy our product over others. If we ever came down to average stud prices, we could always build order files and sell additional product."

The vast majority of Ainsworth's sales were in the U.S. and Canada, with less than 10 per cent — mostly 2×3s — destined for the U.K. Offshore sales were handled by a consortium called Seaboard Lumber Sales Ltd., which dated back to the 1930s.

It was around 1980 when Seaboard felt the timing was right to take another look at the huge Japanese market. Given Ainsworth's reputation for producing a quality product, Seaboard felt their chances were strongest with Ainsworth onboard.

The Japanese were still reluctant to buy lumber products made outside the country, which put the onus on Canadian producers to show they could meet Japan's strict tolerances.

"For Japan, it just has to be quality, quality, quality," says Dick Sellars. "I like to say it took them about five years to train us to do it well enough for them."

It took several overseas trips by Sellars and Seaboard reps to get the program started. Studs destined for Japan had to meet "J-grade" standards, with little allowance for imperfections. If a board's length was out by less than a millimetre — basically the thickness of the end-paint — they'd be sending back complaints that the lumber was off-grade.

Having become familiar with Japan's requirements, Sellars would return to 100 Mile and "make everybody mad" by raising the bar on production and packaging standards that were already among the industry's highest.

"We started the export program off with 100 Mile because they had a slightly better timber supply than Chasm, so we knew we'd have the best material to pick from," says Sellars. "Then we'd show the production people what we wanted. We would have incentive programs for good grading and quarterly awards with plaques. We would really try to impress on

High grading standards were paramount for success.

everyone that we had to make something special for these customers. We wanted to nail that quality down, and when we were able to do it at that level, we'd go down to Chasm and get them going on producing for Japanese grade."

Close attention to lumber grading played a key role in Ainsworth's success in producing premium-quality studs. From early on, David Ainsworth encouraged all employees to get their grading ticket whether or not it was directly related to their job. Those who did were eligible for a slight premium on their hourly wage.

The painstaking process of meeting high standards on a consistent basis was soon paying off. When Ainsworth entered the Japanese stud market in 1980, they were among several lumber producers from western Canada vying for offshore business. By 1982, Ainsworth's studs were ranked number one in quality and brand recognition; they had become the gold standard to which the others were compared.

"The Japanese tolerances were tight and their grades were high, but we got the reputation of being one of the most quality-minded producers in Canada, and we had an order file that did us very well," says Sellars. "So when the prices in the U.S. fell off, we were able to maintain those high prices in Japan."

By 1984, studs produced at Chasm were also being sold in Japan.

Roughly 12 per cent of Ainsworth's total production was now being sold offshore, and of that, 90 per cent went to buyers in Japan. The remainder was sold in Great Britain and several European countries.

With the company's focus on premium-grade studs and their already distinctive turquoise end-paint, the Ainsworths felt it would be advantageous to develop their own brand rather than continue under the collective sales of Mountain Pine Lumber.

The result was an amicable split of Mountain Pine in 1981. Pinette and Therrien kept the Mountain Pine name, while Ainsworth and Weier's Sawmills formed Ainsworth-Weier's

Ainsworth always maintained a number of company-owned logging trucks.

Lumber Sales, which served as a co-op sales office for the two independent companies. Their new partnership would last until 1986, when Herb Weier was ready to retire and sold his company to Slocan Forest Products, which handled its own sales and marketing activities.

Rod Lee was the sales manager of Ainsworth-Weier's at its inception and was joined by new recruit Blair Magnuson. Magnuson had only been with Mountain Pine a few months when the split occurred, but he had already developed an appreciation for how these companies operated.

"The attention to quality at Ainsworth really stood out," says Magnuson. "It was guys like Carl Watson and Dick Sellars at the mill, Eugenio Cargnelutti as the planer foreman — all those guys. Boy, if you went to them and said something was wrong, they were on it right away. They took it like it was theirs."

Meeting the company's founder made an impression as well.

"The first time I met David Ainsworth, he sat me down in the coffee room and told me how important I was to the company," recalls Magnuson. "I couldn't believe the owner of the company sitting down and saying this. It made you feel good."

Regardless of which company was handling the sales, the dominant means of transportation for Ainsworth and the other Interior mills remained shipping by rail. Customers put great stock in a producer's ability to deliver the product on time and in good shape. And since transportation costs were calculated into the selling price, lumber companies paid close attention to the railway service. The government-run PGE (Pacific Great Eastern) became British Columbia Railway in 1972 and BC Rail in 1984.

Shipping a load of lumber from point A to point B wasn't always a smooth ride. Packages inside a poorly loaded rail car that got banged around en route could shift position, making them difficult to unload at the final destination. In extreme cases, loads that had shifted during transit had to be cut with chainsaws, or partially unloaded by hand, just to gain forklift access.

The cost of delivering the product, whether by rail or truck, is factored into lumber's selling price.

"You should have seen some of those bundles; they were a sight to behold," says David Ainsworth. "You couldn't get the damn door open — and that was just the railroad picking them up at the mill."

If a customer complained about a car arriving at its destination with lumber every which way, it often entailed a trip by an Ainsworth rep to sort out the issue.

Ainsworth shippers soon refined their car-loading technique to the point where customers were willing to pay a $2 to $3 premium per thousand for the packing job alone.

One of the more curious features of the transportation scene at the time was the phenomenon of transit sales. Carloads of lumber would be shipped out from B.C. producers to make their way to one of several key transfer centres in the U.S. The lumber had yet to be sold, so the load actually had no final destination when it set out on its journey. Oftentimes these loads were sent by the slowest carrier or the most indirect route available in hopes that salesmen could find a buyer while the order was in transit. It wasn't unusual for such cars to be in transit for a month or longer. If they reached the transfer point, say in Minneapolis, and there was still no buyer, the car would be moved off the tracks until its contents were sold.

Transit sales were of far more benefit to the lumber producers than the railroad companies. For sawmills, it meant low-cost storage space and reduced inventory in the yard. Not surprisingly, the practice was eventually brought to a halt when railroads began to experience car shortages and excess traffic on their lines.

Unfortunately for the producers, railcar shortages didn't end when transit sales were phased out. For as long as lumber companies have been shipping their product by rail, they've

---

It was New Year's Eve day and we had one little job: we had to move a D8 Cat on the old Highway 24. It was minus 25 and it really snowed that day. Robin Nadin was piloting out front in his pickup. Met Kozakevich was driving the low-bed with the D8 and I had the dump truck with the blade for the Cat in the back. There was a bad little hill and Met spun out at the top with the trailer. It was too steep to put on chains, so Robin suggested we jump the Cat off the low-bed and push the low-bed up with the Cat.

Robin started the Cat and backed it up a bit. The low-bed trailer wiggled a bit and away it went. The tires had warmed up a little and the whole thing slid to the side. The low-bed kind of jackknifed and the Cat slid off its side, so there we are with the road blocked and the Cat on its side, and I'm in the dump truck at the bottom of the hill. When I saw this trailer start sliding, I had the truck in reverse and was racing backwards.

Jack Black from Highways came along the road and said, "You guys are the fastest at unloading Cats, but you've got to fine-tune it a little bit — you're supposed to land on the tracks."

The WCB guy came by, looked at us and drove away.

We had a D6 Cat a bit further out so Robin went in his pickup and brought the D6 back. We pushed the low-bed to the top of the hill. Then we hooked onto the overturned D8, turned it right around, then hooked a winch line on to pull it over and up around against the bank so it wouldn't fall or crash.

Then we called Bob Glover in the shop and asked if we should try to start this Cat that's been upside down for over three hours. He said try it and if it jams up, quit. Robin started it up and it turned over. We got the Cat back on the low-bed at the top of the hill and didn't get home until about six that night. And this was all supposed to take a few hours in the morning.

— BOB McCORMACK

had to put up with periodic car shortages that result in delayed deliveries and a crimp in the cash flow.

One option was for a company to lease its own railcars. However, it wasn't only car shortages that could affect shipping; strikes and equipment-related disruptions were also problems. Ainsworth opted for buying highway transport trucks to make it less dependent on rail travel altogether. The company started out in 1975 with two low-bed transport trucks that were on the road constantly.

Dick Dickson, who began driving logging trucks for Ainsworth in the mid-1960s, would make as many as five trips a week to Vancouver, Prince George or Vancouver Island, covering some 8,000 miles a month. Rather than return empty, he would haul back mill equipment and supplies when needed.

By now Ainsworth's truck fleet included eight logging trucks, the two highway trucks, a couple of low-boys for moving heavy equipment from site to site, an assortment of gravel trucks and even a reconditioned cement mixer.

"We used to argue, and I think rightfully so, that it was good business to have some of our own logging trucks; not all of them but some," says David Ainsworth. "Then when we were in a new area, we'd send our own trucks out with drivers we could trust. They could go into a new area and establish the time required to get out there. Then we would build a route or get a price for that particular route."

Long-haul driver Dick Dickson.

Between the Exeter, Chasm and Lone Butte operations, haulers were bringing in 100 loads of logs a day. The goal was to have a backup of approximately 800 loads in a log yard by mid-winter to carry the mill through the breakup period.

Operating its own transport trucks reduced Ainsworth's dependence on rail transportation.

Driver Bob Teichgrab and one of Ainsworth's first trucks converted to seven axles.

Looking after all of this hauling activity was Bob McCormack, who joined the company in 1975 and was quickly put to work on a D8 Cat to develop the log yard at Chasm.

McCormack had been around trucks and heavy equipment all his life and by 1977 was put in charge of dispatching Ainsworth's logging trucks and the trucks of 30 or so contractors hauling to Exeter and Chasm. As the job progressed, McCormack was also involved in Ainsworth's efforts to improve hauling efficiency by modifying the trucks and trailers themselves.

They were able to improve their performance in two ways: either by lengthening the load or lowering the profile of the entire truck and trailer.

Lengthening the load was first achieved by adding a sixth axle to the log-hauling trailer, soon followed by a seventh. Ainsworth also pioneered the use of a trailer attachment called a "dog logger," which allowed for even bigger loads to be brought into the mills.

By lowering the truck's cab and trailer, they were able to lower the rig's centre of gravity, which made for safer hauling.

"The idea was to haul more wood without rolling over, so we put lower chairs in every truck we had. Then we lowered the frames and put on low-profile wheels. We had low-profile bunks, then we stretched them out and added two more axles," says McCormack.

Before any of these changes could be incorporated, the company had to prove to highway authorities that the loads still met the appropriate safety standards. Some of the testing involved chaining a fully loaded truck to a tilt deck to determine at what angle its wheels would leave the ground.

"We spent a lot of time doing that; it was a big job," says McCormack. "We had to prove we could haul longer, lower loads more safely. It was hard to sell something like that. When we would buy a new truck, like a Freightliner, Brian and I would go and tour the factory to get ideas. We'd research those trucks pretty thoroughly before we bought one."

Jack Lefferson changes the saw blades at the Chasm mill.

By 1980, Ainsworth's truckers were using a new $40,000 radio communications system that provided coverage from logging operations near Williams Lake in the north to the Bonaparte area in the south. The new repeater system greatly improved trucker safety on the back roads, not to mention providing a more efficient means of communication for dispatchers like Bob McCormack.

Throughout the late 1950s and early 1960s, Ainsworth Lumber was too small an operation for a labour union. But once there were over 100 hourly employees by the late 1960s, the IWA began to show interest.

Virtually every other mill of equal or greater size in the Interior was now unionized, but Ainsworth's employees chose a different path: in 1970 they voted to form an independent entity called the Cariboo Woodworkers Association. It was a certified union with full bargaining powers.

The decision to go it alone wasn't a difficult one: Ainsworth had traditionally paid wages equal to or slightly above union rates, so the prospect of getting higher wages by joining the IWA didn't apply. In the meantime, relations between employees and management were on a solid enough footing that additional leverage wasn't required when it came time for workers to bargain collectively or resolve disputes.

As it turned out, in 1981 this level of trust between the CWA and Ainsworth Lumber would play a crucial role in the company's survival.

The markets in 1979 were at a record high, but by 1980 they had plunged to their lowest point in over 35 years. Lumber was selling for $120 a thousand board feet. It was the worst global recession in half a century. In an attempt to battle rising inflation, the Canadian and U.S. governments hiked interest rates, and bank lending rates rose to 20 per cent or more. Mortgage rates followed suit, and housing starts in the U.S. dropped precipitously.

The demand for lumber fell accordingly, and B.C.'s producers were in a serious bind.

For a short time, the pulp market remained relatively strong and chip prices were stable. But as Interior mills continued to operate to produce chips as a byproduct of lumber, they only added to an already huge oversupply of lumber heading into the U.S.

With prices so depressed, mills were operating at a loss for months on end. Throughout British Columbia, many operations were forced to curtail production or shut down completely. In November 1981, nearly 7,000 lumber industry workers throughout B.C. were laid off in one week. Meanwhile, lumber production throughout the central Interior had dropped by 26 per cent.

Ainsworth was feeling the crunch like everyone else. Employees had responded to a call for greater cost efficiencies and production increases, but after three months the company's financial situation had only grown more desperate.

To shut down operations until the markets turned around was the least desirable option: in addition to the economic hardship on employees' families and the local economy, the loss of cash flow would lead to Ainsworth defaulting on payments, which could easily result in bankruptcy. Larger companies with cash reserves could survive such a shutdown, but Ainsworth was not one of them.

Their only recourse was to reduce costs even further, and labour was the biggest expenditure. Early in November, Ainsworth approached the CWA executive with a proposal to roll wages back 19 per cent to pre-June 1981 levels — a drop to $10 an hour from the then-current rate of $12. The rollback would also include management's salaries and contractors' rates. The rollback would be reviewed by both parties in eight weeks to decide whether an extension was necessary.

When the controversial proposal was put to a ballot, CWA members were solidly behind it: 80 per cent at Exeter voted to accept, as did 95 per cent at Chasm. In total, the rollback affected 250 CWA employees, 45 managers and office staff, and 85 loggers and truck drivers.

If there were those employees who had lingering doubts about the company's financial straits or the dire state of the industry, they had only to look at their neighbours down Exeter Road. No sooner had CWA members approved the rollback than Weldwood's management announced the indefinite layoff of 125 sawmill and planer mill workers at their 100 Mile operation.

Ainsworth's agreement with the CWA was certainly headline news in the Cariboo, as well as other centres around the region. The local economy was taking a beating, and an indefinite shutdown by one of the area's largest employers was the last thing the community needed.

Even Kamloops MLA Claude Richmond got in on the act. Richmond said in press reports that the Ainsworth rollback agreement "broke new ground in labour-management relations in B.C."

But it didn't end there. Within two weeks this pact between a small Cariboo lumber company and its employees was drawing interest from news outlets around North America — from Vancouver to Toronto to New York.

Ainsworth's controller at the time, Ken Stewart, found his phone was ringing off the hook when the word started to spread.

"When we first looked at the idea of a rollback, we thought it was a great local solution," says Stewart, who left the company in 1985. "It was purely a financial decision. We weren't doing it for publicity so we were quite surprised at the feedback — especially when the TV cameras showed up."

Joe Pacheco, a long-time Ainsworth employee who was CWA president at the time, was also inundated with interview requests from across the country.

The local community would have been hit hard by an Ainsworth shutdown.

100 Mile House in the mid-1980s.

"We didn't need an independent audit of the company's books," Pacheco told one reporter. "Everything was set out in the open and we are satisfied we have the true picture. They are not pulling any wool over our eyes."

Beyond the rollback agreement itself, reporters were also playing up the community spirit angle. A local bakery manager got the ball rolling when he took out a newspaper ad announcing price reductions as a show of support for the Ainsworth workers. Other businesses followed suit, from a building supplies store to a pharmacy, and with Christmas fast approaching, the story gained momentum with each new telling.

As reporter Pamela Martin of BCTV put it, "There is something happening in this community."

Out at the Exeter mill, production actually increased during the rollback. An extra man per shift was hired to fill in during coffee breaks so that, aside from routine maintenance work, the equipment could run virtually non-stop.

By late January 1982, Ainsworth and the CWA met again to review the situation. There had been no improvement in the lumber markets and the prospects for any sudden changes looked dim. Faced with the same choices as before, the CWA workers voted 86 per cent in favour of extending the rollback at least one more month, at which time they'd review the situation again.

As the CWA's Joe Pacheco commented in a follow-up newspaper report: "We're still working and that's the best thing we can do right now for our membership. We don't want to see the company shut down and we don't want to be out of work ourselves."

As it turned out, in the end the CWA employees actually came out ahead financially by agreeing to the rollback. When markets improved and wages returned to pre-rollback levels by early spring, David Ainsworth acknowledged the CWA members' loyalty by adding a special premium to their hourly rates.

In later months when the IWA negotiated a new contract for its Interior members, it included a series of increases based on certain percentages. Since Ainsworth generally matched or exceeded the IWA's rates, the CWA's wages were increasing accordingly, with their additional premium still intact. Before long the CWA members regained all the wages they had lost due to the rollback, and then some.

"We waited a long time to take that premium off," says David Ainsworth, "longer than we actually had to. But it did a lot for the company's credibility."

On the sales and marketing side, Ainsworth's efforts to maintain production while other mills were closing down didn't go unnoticed among its customers.

"It was wonderful to be able to go to our customers and say, 'We're not going to disrupt shipments,'" says Rod Lee. "We said, 'We're going to keep producing for you because we know you still want to buy our products.' Being able to stay in business like this put us in a growth position in the market. That was one of the unique things that came out of the rollbacks."

Ainsworth survived this crisis, but the continued movement of B.C. softwood lumber into slumping U.S. markets set the stage for yet another major stumbling block, this one involving international trade.

By 1982, U.S. lumber producers who were forced to curtail production were looking for a way to slow the flow of surplus B.C. lumber into already depressed markets. Their solution, under the provisions of U.S. trade law, was to impose a countervailing duty on B.C. lumber products.

The B.C. industry fought the duties and the U.S. producers lost this particular dispute, but the battle over B.C. softwood was only beginning. The coming decades would see an exhausting — and expensive — series of charges and counter-charges, negotiations, lawsuits, court rulings, duties, tariffs, decisions from international trade tribunals and countless appeals from both sides of the border.

Closer to home, the events of 1981 indirectly contributed to the closure of Ainsworth's Lone Butte mill, purchased from the McMillan brothers only four years earlier.

Workers at the Lone Butte division were already members of the IWA when Ainsworth acquired the mill, which precluded them from participating in the rollback agreement because of union policies. When the Lone Butte employees went out on strike as part of a wider IWA campaign unrelated to Ainsworth, the company, after several months of waiting, decided to close the mill altogether and redirect the timber supply to its Chasm and Exeter operations.

The planer was eventually sold to an operation in Carrot River, Saskatchewan, and the mill equipment was gradually sold off over the following years.

The Chasm sawmill, meanwhile, was barely six years old, but pressure was mounting to improve its efficiency.

"You either have to keep upgrading or get off the playing field," says Al Smith. "We were at a point where we had to be more efficient and get more production per person."

By early in 1982 the production at Chasm had nearly doubled, from 45 million to 90 million board feet of lumber annually. The company had added a complete new saw line hinging on a Mark II Chip-N-Saw, along with new log-sorting equipment and a new lumber stacker.

"With these changes we were getting more wood into the mill and putting more wood down each line with the same number of people," says Smith.

Ainsworth had grown to 500 employees by the mid-1980s and was enjoying the improved market conditions after a prolonged downturn. But it was still a family-run company with

Long-time Exeter employees around 1985. Back row, from left: Brian Ainsworth, David Ainsworth, Joe Pacheco, Wilbur Cornthwaite, Norm Woods, Pete Nadin, Paul Loeppky, John Tomlinson, Barry Vilac, Dick Sellars, Allen Ainsworth, Phillip Bolha, Jack Lefferson, D. Rohman. Front: Al Smith, Ed Kraft.

just two main sawmill divisions barely an hour apart. Compared to larger operations with layers of middle management, Ainsworth's chain of command was as lean as they come.

David was still company president, but Allen and Brian were now playing lead roles, backed by an able team of managers.

Being relatively small and family owned, Ainsworth Lumber had never placed too much importance on job titles and formal job descriptions. Bob Glover was moved out of the shop and was managing the Exeter plant for over a year before he finally asked if his new duties were permanent.

"I started managing the mill site, and a year and a half must have gone by when I saw Al Smith one day and said, 'Am I doing fine or do you want me to go back to the shop?' Al said, 'No, just stay right there.' That's how they approached it; if you were doing a good job they just left you alone."

Nor did the Ainsworths' management style leave a lot of room for over-analyzing the pros and cons of a particular course of action. As David would say, "The only time you stand around with your mouth open is when you think you're going to learn something."

The depth of knowledge among Ainsworth's various foremen and managers, most of whom had been with the company for decades, was one of the company's greatest strengths. Former controller Ken Stewart was the first employee with a university degree the company had ever hired.

"One of the interesting things about Ainsworth at this time was the number of senior managers who had worked their way up through the company," says Stewart. "It was rather a unique company because you didn't have the executive-type groups. Everyone came up through the ranks: Dick Sellars, Al Smith, Bob McCormack, Bob Glover. They were all guys who knew their stuff. Not everyone there was polished, and they didn't try to be polished, but they knew how to run things and that's what they did. That's where the success came from, that was the backbone. You'd get people who were extremely talented and knowledgeable about their aspect of the business. You put them all together and it worked out quite nicely."

Anyone who enjoyed working without a lot of oversight was in the right place.

Contractor Ken Greenall recalls that for some construction projects Allen Ainsworth would just rough out a sketch in the dust on the hood of his truck. After a minimum of discussion he'd drive off — with the plan still etched on his hood — and leave Greenall to carry on building.

Allen was famous for saying, "If I have to show you how to do it, I might as well do it myself."

On other occasions, a foreman would get Allen's input on a particular issue and proceed accordingly. Then Brian Ainsworth would come along and have a different idea. Managers quickly learned to act on the most recent set of orders, and there were never any repercussions.

If Brian saw someone lingering at the office, he'd say, "If you don't have anything to do, do it somewhere else."

When the company hired Steve Silveira in 1980, he was the first "outsider" to join Ainsworth at the senior management level. He also had a university degree in forestry. Brian Ainsworth was looking for someone to head up the woodlands department — with a minimum of supervision — and report to him directly. Silveira had been a logging superintendent at Weldwood's 100 Mile division, and Brian had heard positive reports about his performance.

The job was a huge leap in responsibility for Silveira and he started out with a good deal of trepidation. He was only 32 at the time and many of the men under him were 10 years his senior. They were the same seasoned, long-term employees who had provided Ainsworth with a depth of experience over the past two decades.

It was spring breakup and Brian said, "We've got to have a D8 Cat at Chasm right away." Well, the only Cat I had that I could get there was in Williams Lake, where it had just been rebuilt. You're not allowed to haul machinery while the road restrictions are on, so Larry Duncan and I went up to Williams Lake at midnight and loaded this D8 onto a low-bed. Then we came home the back way through Dog Creek. Pulling out of Williams Lake and here comes a police car up behind us and it's midnight and we've got this Cat on a low-bed. There was no point in taking the blade off because we were illegal anyway, so we might as well be wide and illegal. It's only another couple of hundred bucks in fines. So we're sure this police car is after us, but he went right by and never even looked at us.

On that 1100 road back there, you're never even sure where you are. We'd have to jump the Cat off and pull the truck through some of the mud holes, then jump the Cat back on and carry on down the road for a ways, then get stuck again. Then we'd have to take the Cat off and plough the snow where it had drifted in. The trip took us all night. We got to Chasm at eight the next morning. You'd never attempt something like that now. They'd seize your Cat and the low-bed, and you'd get major fines. That was the last time we did that. It's the midnight express, and away you go. None of the Ainsworths knew anything about this. Like Brian always said, "It's easier to get forgiveness than permission."

— Robin Nadin

Ainsworth employees at an eighties Christmas party are entertained by The Pigeon Creek Turkeys, otherwise known as the woodlands crew. From left: Robin Nadin, Bob McCormack, Bob Phaneuf, Omer Gosselin, Ed Kraft, Hanns Fellenz, Doug White, Luigi Sposato, Don Brown, Steve Silveira.

To make matters even more interesting, Silveira learned soon enough that David Ainsworth had his own opinions about how this university-educated newcomer should be conducting business on behalf of the company.

"My first introduction to David was when he came into the office one day and said, 'Young man, I need to have a chat with you. Come out to my pickup.' This ended up happening on several occasions, and the talks were pretty consistent," says Silveira. "I started to call them 'my front seat lessons.' He would always start out by saying a few words, and then basically he would tell me that he thought my education was getting in my way."

Silveira was defensive at first, but it didn't take long before he saw things in a different light.

"As I learned more and more about David, I came to respect this ability he had. He could size things up quickly and from a very practical point of view. He could make decisions without a lot of analysis. I tended to be more analytical and I wanted to be sure everything was right."

Silveira saw firsthand how the company was able to make quick decisions while he was still involved with purchase wood at Weldwood. Weldwood was considering buying a sizeable block of private timber near Mahood Lake that would cost around $300,000.

"It was a good deal, very good timber, so the company was grinding through the process of getting all the approvals from head office in New York and back again. In the meantime, Brian Ainsworth got word of the timber, went out there and wrote a cheque to the owners and scooped it out from underneath us."

Silveira took his edict from Brian to "manage the business" seriously, and he did ruffle some feathers early in the game. One such occasion involved his push to improve the company's utilization of its timber by bringing smaller wood into the mill. Those in charge of production at the mill weren't too happy about it and let him know.

In due course he'd find himself in David's pickup again for another lesson.

"It's not that David would always disagree with me, but he would caution me, wondering if I was taking advantage of all the experience of these guys here."

Steve Silveira wasn't the only new addition to the management ranks.

The 1980s also saw David and Susan's daughter Catherine taking on increased responsibilities within the organization. Catherine had been a fixture at the Exeter office throughout her infancy and childhood — spending time in a crib in the corner or a playhouse in the woods — and then progressed to summer jobs from there. By the age of 20 she was ready to graduate from college in the Lower Mainland and had no plans to work for the family business.

A phone call from Susan changed all that.

The company office had recently entered the computer age and some new equipment was being delivered. Susan needed someone to learn the ropes in short order and Catherine was her best prospect.

"That was my first introduction to anything that had to do with computers," says Catherine. "It wasn't really a matter of being interested in them. They were just a tool that you use to get the job done."

Catherine's full-time presence at the office was an immediate help for Susan, who now had some direct backup for a growing number of responsibilities. Her initial jobs were mundane by any standards, but it didn't take long before she was providing input on modernizing various procedures or equipment around the office. Her expanding expertise on computers and their best application in the office was entirely self-taught.

> *Catherine had to learn finance and accounting the hard way: on the job. She had no accounting training, but today, for all intents and purposes, she probably knows as much as any accountant around this place. She knows computers too. When a younger member of the family comes into the business, they have to find where they fit. There's a lot of shoes to fill around here, and some of the shoes still have somebody in them.*
>
> *So when Catherine came into the business, computers were starting to be used. Susan, David, Allen and Brian were too busy doing other things, so there was a wonderful opportunity for Catherine to really dig in on the office and accounting side of the business. Catherine had everything to do with computerizing the company in the Exeter office days. I remember watching her; she was literally helping to program computers.*
>
> *She brought that self-taught expertise to the company.*
>
> — MICHAEL AINSWORTH

Catherine Ainsworth joined the company full-time around 1980.

"Joining in on decisions was a gradual thing," says Catherine. "It's a natural phenomenon in a family company that those who have an inclination to manage or to lead simply do. It's not like you get a promotion or someone makes an announcement. It's more like there's work to be done and you do it. Then, once you are doing that sort of work — if you've got the knowledge and expertise to manage a certain part of the process — people call on you for opinions or include you in the decision-making process. It's not something that you can put your finger on. It's not even informal — it's more of an evolution."

Catherine married university student Hanns Fellenz on June 20, 1980. Hanns had joined the company the previous summer, learning the ropes from the ground up.

"The very first job I got was at the Exeter mill working for Al Smith," says Hanns. "I was put to work shovelling gravel into the base of one of the transformer stations above the mill. I was pretty green and from the city — not used to manual labour — and Al Smith certainly put me through my paces. I was hurting, but I survived."

Hanns went back to university in the early 1980s and returned to Ainsworth several years later, initially working with the forestry crew.

"I was working with the group that was out there doing the timber cruising, and they were all pretty fit. So you had to keep up with those guys — otherwise you'd be left behind with the bears. Operating snowmobiles was an experience, too. They never seemed to run an entire day without breaking down."

> *I don't know why, but our first computer systems were called "Arnold." Arnold-I was so bad that David Ainsworth drove it to Vancouver in a pickup truck and dumped it on the supplier's loading dock. Arnold-II had 128 kilobytes of RAM, and each removable disk — 18 inches in diameter and weighing 20 pounds — held an incredible 5 megabytes of data.*
>
> — HANNS FELLENZ

Within a short time, Hanns was also developing expertise in computer technology and, under Pete Nadin, was building a database for use in creating five-year forestry development plans.

"This was my first foray into computers, and from there I got involved in other projects with computers."

By 1987 he had joined Exco Industries, bringing the company into the computer age with the introduction of computer-aided design and drafting.

Since the early 1990s, when he left Exco as general manager, Hanns's primary responsibilities with Ainsworth have been in the field of information technology, the area he currently oversees.

Although Catherine Ainsworth was the sister of Allen and Brian, she was actually closer in age to their sons — Michael, Douglas and Kevin — who were now emerging on the family business scene.

Like their fathers before them, the three boys had grown up around the lumber business from day one — minus the cabins, cookhouse and portable sawmill. With their mothers working for the company when they were young, the boys thought of the old Exeter office as their second home.

Later, if they weren't tagging along with Allen and Brian — sitting in on meetings at the Tastee Freez diner or the Red Coach Inn — they were up at the Exeter site, playing in one of the old family cabins now sitting empty. The boys also had plenty of opportunity to ride shotgun on the grader as Allen levelled the entry road to the mill.

By age 10 or 11 they were racing around the Exeter site area on their dirt bikes, and a couple of years later they were enlisted in their first summer jobs piling planer strips, shovelling sawdust or sweeping up on the cleanup crew.

Allen Ainsworth (left), the three boys and Ken Greenall set down at an area lake.

"At times we behaved badly," says Kevin, "but for a 13-year-old it was good experience."

On occasion, the boys were also allowed to drive a company logging truck or move a few logs with a 966 loader — one of the highlights of a Saturday visit to the Exeter or Lone Butte site with Brian or Allen.

The summer jobs continued through the high-school years, whether the boys were timber cruising with Doug White, Bob Phaneuf and Ed Kraft or, as in Michael's case, operating a loader for a season at a company-run placer mining operation near Quesnel.

Being the eldest of the three, Michael was the first to set off for university in Vancouver in 1981. If his parents or grandparents had hopes of him returning after graduation to work for the family business, it never came up in conversation. The door to other careers was left wide open.

"It was always assumed we were going to university; there was never any question in my mind," says Michael. "That's what you did after high school. But there was never a discussion about what you were supposed to take or what you would do with it afterwards."

> *When I was in my late teens, I used to have talks with my grandmother in her office. She'd be opening the mail or something. I wouldn't say they were lectures, but more like pep talks on running a small business. She'd give me her views on running a small family company, running it responsibly and watching costs, understanding your business and other things about suppliers. She was a tough administrator and a pretty savvy small business operator, so she tried to impart some of her values to me.*
>
> — Kevin Ainsworth

Based on his summer mining experiences, Michael decided to concentrate on the sciences, thinking that geology would be a good fit. By the end of his second year, however, he was having serious second thoughts. His summers were still spent working for Ainsworth, whether it was answering the phone and typing up bills of lading at the Exeter office or working at the lumber sales desk in Vancouver.

It was only after a year-long break that Michael came to some important conclusions about his future. The time away from school included a stint working for the company back in 100 Mile, as well as backpacking around Europe for several months with his cousin Kevin.

"That trip to Europe was a real watershed for me," says Michael. "We had no itinerary or plans and just figured it out when we got there. I was really on my own for the first time, and when I got back — that's when my ideas crystallized: I'm not going to work somewhere else, and I have no interest in geology. That's when I started digging around and said maybe I should look at the faculty of forestry."

His decision was a step in the right direction, but there were still some turns to make. The more he looked at forestry and its emphasis on resource management, the more he realized it still wasn't the right fit.

"The resource management side didn't necessarily appeal to me because all of my experience was with Allen on the milling and marketing side. Had I spent more time with Brian, maybe I would have focused on the resource management side, but that wasn't the case."

Still unsure of which avenue to pursue, Michael came across UBC's wood science program, and the search was over.

"I discovered that this program was all about wood chemistry, manufacturing, timber engineering and timber design, and it all seemed fascinating. I got into that and dug in hard, and school seemed so much easier because it was all more interesting and relevant."

Since 25 to 30 per cent of the course was optional, Michael chose subjects from the faculty of commerce to round out his wood science degree.

"At the same time I was learning the technical aspects of wood drying, milling, pulping and wood chemistry, I was also getting training in accounting, marketing, transportation, finance and economics. I couldn't have designed a better program for myself."

Michael graduated with his wood science degree in 1986, but it would be several more years before he was with the company on a full-time basis. Along the way was another degree, this one a master's in forest products marketing, and two years of research work for Robert Fouquet at the Council of Forest Industries (COFI).

"It was the head of the wood science department, Dr. David Barrett, who first introduced me to Robert," says Michael. "Dave had approached me about my plans for the summer. He needed a student to work with Robert on a research project."

Fouquet would eventually join Ainsworth in 1989 and go on to become vice-president of marketing.

Like his brother Michael, Douglas Ainsworth never felt any pressure to join the family business when it came time to choose a career. As far as he was concerned, though, his staying on was a given.

"I had always intended to work in the family business," says Douglas. "I never really considered doing something else. I enjoyed working with my family, and as long as there was a family business, I intended to work there in any capacity."

Throughout his summer jobs with Ainsworth, Douglas had always preferred the hands-on

> *When David and Susan were living down at their duplex, Douglas and I used to have to walk right by their house on the way to elementary school at the other end of town. We'd dawdle and fiddle around the whole way down to school, and we'd stop in at their place and fool around there long enough until we were going to be late. It was a big game we played so Susan would have to give us a ride. We'd hang around long enough, stop in and visit, and then she'd have to load us in the car and drive us literally the last 200 yards.*
>
> *On the way home in the afternoon she usually wasn't at home, but we'd stop in almost every day and peek in, and if she was home we'd stop. Otherwise we'd just carry on and walk home.*
>
> — MICHAEL AINSWORTH

Kevin Ainsworth was on hand when a truckload of new skidders arrived.

Ken Greenall and Michael Ainsworth.

Douglas Ainsworth started out at Exco Industries in 100 Mile House.

mechanical type of activity. Working with knowledgeable millwrights like Hal Young and his crew installing equipment and learning how to weld and cut steel was time well spent.

Not surprisingly, Douglas's entry into full-time work at Ainsworth began at 100 Mile's Exco Industries around 1989, where his first order of business was learning computer-aided design and drafting. By 1991 he was managing Exco before moving on to other projects.

The youngest of the trio, Kevin Ainsworth, took an interest in the forestry side of operations, no doubt kindled by his summer jobs on the woodlands crew, as well as by his father's role in that area. Kevin started out timber cruising, and by Grade 12 he was working with Don Brown, surveying roads and new logging areas east of 100 Mile.

Studying forestry at UBC with his cousin Michael seemed a natural choice, although he initially entertained thoughts of studying marine biology.

"It sounds funny but there was no pressure about staying in the family business," says Kevin. "It just wasn't mentioned. Growing up, we were very focused on our day-to-day activities. But we eventually developed our own interests as we went along. By the time Michael and I were in the forestry program, we were very gung-ho to work in the company."

> Allen had a gravel runway at Chasm so he could fly in there with his Cessna 185. I was there with my cousin Kim Blacklock — we were teenagers — and Allen pointed at this runway, then he showed us the end of his thumb and said, "Anything bigger than that I want off the runway." He wanted us to pick rocks off the runway, and at that point it looked like it was a mile long. So we're looking down what's probably a half-mile gravel strip and thinking, "You've got to be kidding." We had some rakes and we'd work away at it for a while and then say, "This is pure insanity. We could be here until the end of time." I guess he thought it was one of those tasks that would keep us busy for a while.
>
> — MICHAEL AINSWORTH

Summer jobs during university included time on the lumber sales desk, as well as a stint rebuilding the finger-joining mill at its Lower Mainland site in 1986.

After graduation, Kevin joined Ainsworth's new woodlands crew in Lillooet, working with Lillooet staff surveying mainline logging roads in the rugged Bridge River area. By the late 1980s he was studying with Michael in Vancouver as they prepared for their MBA entrance exams at UBC.

"This was the first time that anyone had ever put any pressure on me to work for the company," says Kevin. "My granddad had a talk with me just before I went into the MBA program. He said, 'I guess you're going back to school for two more years, and that's a good thing, but someone's got to come back to work around here. You can't go to school forever.' It was a good pep talk."

Fresh out of university with his MBA, Kevin returned to the Exeter office in 100 Mile to start full-time as a woodlands accountant.

By the autumn of 1985, Ainsworth was taking steps to establish its first production division outside the Cariboo. The finger-joining plant at Exeter was producing lumber, but the colder temperatures were either hampering production or shutting it down altogether. And transportation-wise, it was inefficient to ship the trim blocks north from Chasm to 100 Mile, then back south again in the form of finger-joined lumber.

There was little choice but to relocate. Allen Ainsworth soon found a former steel-fabricating shop sitting on five acres near Abbotsford in the Fraser Valley. It was so close to the U.S. border that a person could literally stand on the property with a foot in either country.

The first step in readying the site was to bring in truckloads of rock and gravel to firm

CABOOSE PROJECT NEARLY GOES OFF THE RAILS

It was 1985 and David Ainsworth had been asked by local developer Bridge Creek if his men could transport a caboose from Exeter Station to a shopping centre parking lot a few miles down Exeter Road. The caboose had been mothballed by BC Rail and was going to be renovated and used as a store.

The caboose turned out to be far heavier than anyone guessed. It took three 966 loaders over four hours to maneuver it onto a low-bed truck. And that was after the wheels were removed.

The trip down Exeter Road took an hour, and then the real work began. The caboose had to be moved off the low-bed and onto its wheels. Then the whole works had to be set on a short set of tracks already in place.

Right-of-way foreman Robin Nadin was in charge of the weekend project, but David was also on the scene. No one there had ever moved a caboose before, so everyone had an opinion on how it should be done.

"Between me trying to organize some of the people and David trying to organize me or the people there, it was getting kind of exciting," says Nadin.

No one at Ainsworth had ever moved a railcar.

"There was a difference of opinion. David was scaring the hell out of me because he was under there poking the blocks around, and it's just held up by machines. I had in my mind how I wanted to do this, but I'm sure David had a plan too — and they weren't exactly the same. So we kind of had to straighten that out. More than likely he won. It might have been a career decision if he hadn't."

---

up the swampy ground. Inside the building, little modification was required beyond moving a few walls around and judiciously cutting some trusses. Ainsworth electrician John Classen spent the winter wiring the building and putting in power, and in April 1986 they pulled the plug on the finger-joining plant at Exeter.

There was no time wasted on dismantling the plant.

"We had precut some of the beams and legs," says Don McGladdery, "so as they were running the last of the material through the machinery, the construction crew and electricians were coming along and cutting the rest of the legs. The wiring was all numbered and labelled so it went with each piece of equipment. In approximately eight hours we had the equipment removed and loaded on trucks."

The lone exception was the shaper table, which had to be extended, but within a few days it was also on its way south. An Ainsworth construction crew headed up by Hal Young had the mill reassembled and running again in six weeks. They also installed an additional trim line that could precision end-trim natural studs to lengths from five feet to eight feet.

The Fraser Valley's rural character proved beneficial when it came time to hire the 25 new employees at the mill. Work at the finger-joining plant was very demanding because of the repetition and fast pace. Residents from around the area who hired on were well-suited for the job — strong and sturdy with a work ethic to match.

Long-time Ainsworth employees Steve Peebles and Len Teichrob were also there at startup. Don and Sue McGladdery, meanwhile, moved down from 100 Mile and Don became Abbotsford's site manager within several months.

Relocating the plant to a more temperate climate had immediate advantages. They were now able to use a cold-set type of glue that didn't require any thermal curing. It took longer to harden, which provided more curing time, and was easier to manage and clean up. Sue McGladdery's expertise in the glue department played an important part in the ramp-up process, and in the years that followed she would look after quality control for the plant.

Although no one there had worked in a plant like this before, within six months the Abbotsford crews were exceeding 100 Mile's output by at least 10,000 board feet per shift.

The old Evans lumber mill at Savona. At the time of purchase, logs were stored in Kamloops Lake.

By 1987 the lumber mill was 30 years old and was not operated full-time.

Within a year they installed a chipper for the waste blocks that formerly were dumped into huge piles outside the mill. This improved recovery even further. Now the site was less cluttered and they were able to sell material that used to be discarded.

The Abbotsford plant had been running less than a year when the Ainsworths started looking seriously at further expansion beyond the Cariboo.

Oregon-based Evans Forest Products had been operating in B.C. for nearly 20 years, but its parent company in the U.S. declared Chapter 11 bankruptcy in 1986. Part of the fallout included receivers forcing Evans to sell off its Canadian holdings. The company had four production divisions in B.C., two of which were of particular interest to Ainsworth: operations at Savona and Lillooet, both located about two hours from 100 Mile.

Savona ran a specialty overlay plywood mill and a small stud mill, while Lillooet was home to a dimension lumber mill and planer, along with a veneer mill. With the Savona division came two timber licences for a total of 225,000 cubic metres of wood annually. One licence was in the Kamloops Timber Supply Area, the second in the 100 Mile House Timber Supply Area. A couple hours northwest of Savona, in Lillooet, there was a timber licence for an Annual Allowable Cut of 450,000 cubic metres.

> *One of the most interesting things I ever did involved David's old original sawmill. Back around 1980 we bought it back from Louis Judson, who had it out at the far end of Canim Lake. Before he would turn it over to us, he said he had a yard full of wood that he had to run, so Michael, Kevin and I went out. Michael stayed a few days, Kevin a bit longer, and I stayed a full two weeks. Louis showed me how to run the carriage. It wasn't rocket science but there was an art to it. The exciting thing for me was that it was the original mill my grandfather started the business with. Now I had the opportunity to operate it. The mill ran on some basic principles but it was really interesting.*
>
> — DOUGLAS AINSWORTH

The Evans operation at Lillooet included an underperforming sawmill and a veneer plant.

Evans had moved into the B.C. Interior in the late 1960s with the purchase of Savona Timber on Kamloops Lake and Commercial Lumber in Lillooet, along with a plywood mill in Golden and a sawmill in Donald. Between the Savona and Lillooet operations — known as Evans' western division — there were some 370 employees.

Savona's sawmill was originally built to process large fir logs but the supply had long since run out. By 1987 the 30-year-old mill wasn't even producing full-time. Evans had recently made a bare-bones attempt to convert it to a stud mill in an effort to make use of the area's lodgepole pine, but its output remained marginal.

The mill site was fairly unusual in that it had no log yard; the wood was dumped into Kamloops Lake upon delivery to the mill. And if the booming area around the mill became congested, the logs were stored on the north side of the lake and floated over as needed.

Savona's log costs were lower than average, but there was no guarantee of a long-term timber supply. In fact, Evans had already been selling portions of its allotted timber to Ainsworth for several years. Their timber supply areas in the Hihium (pronounced "hi-hume") region were adjacent to each other. The Evans cutblocks purchased by Ainsworth were harvested by Ainsworth, with the timber being sent to the Chasm stud mill.

Savona's original plywood plant, constructed in the early 1950s, was well past its prime and relied on labour-intensive processes. It had long since become uncompetitive in the commodity sheathing markets and was now producing specialty overlay panels — under the Pourform 107 brand — used for concrete-forming systems in industrial applications. Despite the poor condition of the mill, Savona's overlay plywood enjoyed an excellent reputation in the marketplace.

The Lillooet division, meanwhile, had its own issues.

The dimension lumber mill on the site, built in the late 1960s, was designed with large-diameter logs in mind. When Evans added a lathe and veneer mill to their operation in the late 1970s, the bigger wood was being redirected away from the sawmill and to the veneer

Lillooet's production challenges included log costs that were among the highest in the B.C. Interior.

Green veneer produced at Lillooet was sent to the Savona plywood plant.

lathe. The sawmill's head-rig line was now being under-utilized and wood recovery suffered accordingly.

Adding to Lillooet's production woes was the high cost of logs. Much of the terrain in the Lillooet operating area is mountainous and steep — not unlike logging conditions at the coast — with narrow valleys barely wide enough for a creek bed. Every facet of logging activity, from timber cruising and road building to skidding and hauling the logs, is more expensive here than just about anywhere else in the province.

"All in all, it was a very distressed situation," says Allen Ainsworth. "Nobody else in the industry was interested in any part of it. Other operators took one look at it and said no thanks. Evans was losing $30 million a year for the two years prior. So we paid $11 million for the assets, which was relatively inexpensive. Additionally, however, we ended up having to buy $7 million worth of logs, lumber and plywood inventory, some of which had little value."

Ainsworth took over the mills in the spring of 1987, but it would be several more months until the deal was officially closed. There were 22 banks taking part on the Evans side of the negotiations, which also involved a management-led group of employees intent on purchasing the Evans operations in Golden and Donald.

As drawn out as these meetings were, they did give Michael Ainsworth a front-row seat from which to observe the realities of corporate deal making — with international players, no less. Having recently graduated from the wood science program at UBC, Michael spent much of that season with Allen while the Evans deal was being hammered out.

"I would be sitting in on negotiations over at the lawyers' offices for hours and hours. Everyone had cast-iron asses over there by the end of this. Our side of it was pretty straightforward, but what took so long was the management group's side of the negotiations. All in all, it was great experience, though."

Allen Ainsworth, meanwhile, says he "wore out a set of shoe leather on Howe Street walking back and forth between law offices and banks while we were grinding the deal out for seven and a half months."

The negotiations also involved Vancouver lawyers Morley Koffman and Doug Side. This marked the first extensive piece of business that Ainsworth did through Koffman's law offices.

It took two months for Ainsworth crews to dismantle the old Weldwood plywood plant at 100 Mile.

One of the 50 truckloads that transported the components from 100 Mile to Savona.

The re-erected plywood plant was up and running within 10 months.

"It was at this point that we got so much very good business advice from Morley, who has since become a senior advisor and remains on our original board of directors," says Allen.

With the Evans sale finalized, the Ainsworths had once again stuck their necks out on a deal that held no appeal for any of the bigger players. There were some distinct advantages, however, for the self-described "leftover kings."

Their company had now doubled in size. They had forest operations that stretched uninterrupted from 100 Mile House in the north to Lillooet in the south. With plants in relatively close proximity, they could make optimal use of their logs by moving them back and forth depending on the profile of the logs and their end use. The Lillooet timber licences did stipulate, however, that their annual allotted volume of wood, or its equivalent, had to be utilized at the Lillooet operation.

The Evans acquisitions also provided Ainsworth with a more diverse product line. Until this point, all their efforts had been concentrated solely on studs. Diversifying had always been on the family's agenda, but a suitable — and affordable — opportunity had yet to present itself. These new divisions offered entry into markets for green veneer sales and specialty plywood, as well as dimension lumber.

The purchase even had implications for the newly relocated finger-joining plant at Abbotsford. The operation there could now count on a third in-house supplier of trim blocks, the plant's basic raw material.

New ownership, of course, generally spells change, and the Evans acquisitions were no exception. In this case, Ainsworth had to balance the company's needs and desires with those of the provincial government and the IWA, both of which ranked employment concerns high. The Ministry of Forests connected cutting rights to job stability as well.

Once all the parties involved were in agreement, Ainsworth was allowed to close down the stud mill at Savona. It made little sense to operate an out-of-date mill part-time with a limited timber supply that was needed at the more modern Chasm facility.

The plywood side was a different matter. Ainsworth would not only maintain production, but expand Savona's capabilities. This showed an obvious long-term commitment to the Savona operation, as well as Lillooet's veneer plant.

Savona's old plywood mill wasn't worth expanding, and building a brand new facility wasn't an option. So Ainsworth did the next best thing. Al Smith, Hal Young and his crew, and Exco's engineers had plenty of experience by this time dismantling used equipment and trucking it back for installation at Exeter or Chasm. It just so happened that the Weldwood division in 100 Mile had a former plywood plant that had been sitting empty for several years following the upgrade of their sawmill.

Ainsworth bought the building for $50,000 and began dismantling the 72,000-square-foot structure in the fall of 1988. Two months and 50 truckloads later, the re-erection work at Savona began. The plant was upgraded to satisfy new building codes and included extra bracing, new decking and new turquoise cladding. An additional 20,000-square-foot structure for the dryers was also built at the time.

"It was a marriage of the old equipment from the original mill and other second-hand equipment that was purchased," says plywood veteran Jack Duckworth, who was site manager at the time. "We bought a second-hand dryer from the southern U.S., a second-hand hot press from Tacoma and second-hand glue-mixing equipment at an auction. Ainsworth was good at doing this kind of thing. It took about 8 to 10 months to get it all up and running. The transition was slow, though, because we were using both buildings for a time before everything went into the one plant."

Over at Ainsworth's newly acquired Lillooet division, the company decided to keep both the dimension sawmill and the veneer plant operational. Evans had been concentrating on the export market for its random-length lumber up to 20 feet, but with limited success.

"It was such a diversified resource for lumber at Lillooet," says Jack Duckworth. "At the west end of the area you've got coastal-type wood that is green and moist and well grown. In the east it's all dry-belt wood: fir, hemlock, spruce, balsam and pine. It's all kinds of species and growing types."

Even sorting and storing such a variety of wood requires more handling than would normally be the case, all of which adds to the cost of production.

The challenge was clear: they had an underperforming sawmill, unusually high log costs and a veneer mill supplying a sister plywood plant two hours away. The sawmill generally lost money, and the veneer mill usually made money. If Savona bought Lillooet's veneer at market prices and the market happened to be high, Savona lost money. If veneer prices dipped, Savona made money and Lillooet lost money. And around it went.

Underlying this was yet another challenge: integrating the cultures of Evans and Ainsworth to a point where everyone was paddling in the same direction. It didn't happen overnight, and there were some rough patches along the way.

The differences between the two companies were apparent from day one. The number of managers and supervisors employed by publicly owned Evans was in direct contrast to Ainsworth's lean structure with family members at the centre. As a result, some 30 office and managerial positions were eliminated between Savona and Lillooet.

Conflict over lumber quality was just one example — albeit a crucial one — of further takeover difficulties. Evans employees had been trained to push the grade rules to the maximum, shipping out as much off-grade lumber as the rules would permit — roughly five per cent of the total. Being the industry leader in premium-quality studs, Ainsworth's approach was the exact opposite. Their production of off-grade lumber barely rose above two per cent, if that. Although such high standards had an impact on their recovery figures, the

A company picnic for the Savona employees.

In 1990 Exco Industries obtained exclusive North American rights to manufacture the Lokomo, a Finnish-design felling head used on feller bunchers for cutting down trees. Exco became part of the Ainsworth fold in the mid-1970s with the initial design and construction of the Chasm sawmill. The company has since made a name for itself as one of the leading designers and manufacturers of log-handling systems for mills around North America. Its patented single-log feeder has drawn customers from New Zealand, Australia and several countries in Europe. Exco's team of designers and steel fabricators continues to play a valuable role in providing equipment for a wide variety of capital projects throughout the company.

goal had always been to produce the best product possible. Clearly, what was "good enough" in the past at Lillooet was no longer acceptable.

The clash of cultures subsided with time, and numerous Evans staffers turned out to be as loyal as any Ainsworth employee who had been with the company for decades. Along with general manager Jack Duckworth, other Evans staffers who eventually became long-time Ainsworth employees included Gordie White, Bill Davidson, Mickey Forman, Albert Morrissette, Ron Friesen, Ken Viette, Frank Mori, Paul Alain, Marg Lampman, Boyd Shanks, Ed Moroz and John LaRue.

Given the terrain and logging conditions, Lillooet's high overhead continued to be a problem, though. And various developments beyond the company's control only served to squeeze operations further.

The trade dispute with the U.S. lumber industry was rearing its head again, and B.C. policy, determined by the Social Credit government of Bill Vander Zalm, made a difficult situation even worse. The end result was a federal export tax on Canadian lumber shipments to the U.S., followed by a new provincial stumpage system — and higher stumpage fees.

On a regional level, the Ministry of Forests reduced the amount of timber that Ainsworth was allowed to cut from its Lillooet licence area. The total dropped from 450,000 cubic metres down to 348,000 cubic metres annually. Just a decade earlier, the area's Annual Allowable Cut had been as high as 800,000 cubic metres.

Environmental groups were beginning to press the provincial government for an increase in the amount of land devoted to parks and protected areas for wildlife habitat. In the late 1980s the bulk of their activities were still focused on coastal issues, but old-growth areas northwest of Lillooet such as Spruce Lake and the Stein Valley in the south were coming under increasing scrutiny.

Pourform panels were a specialty product manufactured at Savona.

Adding to the mix was the fact that there were seven First Nations communities in the region. The mill itself sat on land that straddled two reserves. It would be a number of years before the area bands would seek a larger role in land use decisions, but their presence was already being felt.

With its high overhead, Lillooet's sawmill was usually losing money. Various equipment modifications were made along the way, however, in an attempt to improve recovery, including a new small-log sorting system. By 1990 the plant was converted to a stud mill. This facilitated the use of smaller-diameter logs, which allowed more of the larger logs to be used as peelers on the veneer side.

"When you run a dimension mill, you're always competing for the same logs the plywood plant requires," says Steve Silveira. "We were always torn between trying to keep the dimension mill going and the plywood plant going. There was far more margin in plywood. The veneer markets in 1987, 1988 and 1989 were very good and the company made a lot of money selling green veneer. With the veneer side being able to sell into the U.S. and to Savona to make plywood, it was a no-brainer."

For a company the size of Ainsworth, the Evans acquisition was a huge undertaking. Ainsworth had grown to five divisions with nearly 700 employees, a sales office in Vancouver and $100 million in annual revenues.

Running along a parallel track, there were developments afoot that were now gathering momentum — developments that would alter the company's trajectory far beyond anyone's expectations.

Although David was still very much on the scene and remained president of Ainsworth Lumber, he was in his late sixties and his direct participation in managing the company had been slowly winding down over the years.

Or so it would appear. After all, piloting Ainsworth's Citation jet was a career in itself that naturally kept him from most day-to-day concerns.

Despite his gradual letting go of the reins, David still lived and breathed Ainsworth Lumber. His keen interest in new technology hadn't faded one bit. Nor had his desire to make the best possible use of the resources at hand.

MAKING THE GRADE

In fact, since 1985 he had been looking into a new type of wood panel that was beginning to gain acceptance in residential construction. The product, made largely from waste wood and conceived as a substitute for plywood, was called oriented strand board — OSB for short.

Where this initial exploration into OSB would lead was anyone's guess. But with half a century of innovation and plenty of risk-taking under his belt, David was still thinking about the company's future. And his instincts were now telling him that OSB was something they should seriously pursue.

It was a bold idea, and it was pure David Ainsworth.

The mid-1980s also saw Ainsworth become involved in the development of the Mt. Timothy Ski Hill

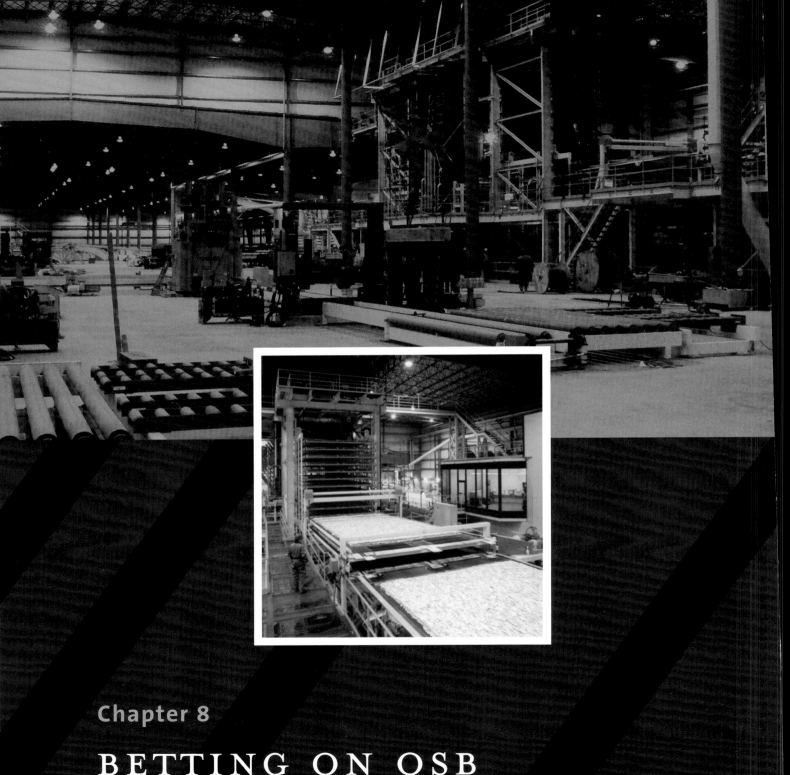

Chapter 8

BETTING ON OSB

Al Owen (right), one of the earliest producers of OSB, tours former Prime Minister Joe Clark through his new mill in Drayton Valley, Alberta.

Al Owen wouldn't call himself "the father of oriented strand board," but he was on the ground floor as far as the industry in North America is concerned. And it was his new OSB plant in Drayton Valley, Alberta, that caught David Ainsworth's attention in the mid-1980s.

David had already been keeping an eye on OSB development thanks to the efforts of former B.C. forests minister, Ray Williston. Williston had been campaigning for government and industry to establish an OSB mill at the coast. It was the minister's foresight years earlier that resulted in the use of Interior sawmill waste to manufacture wood chips for B.C.'s pulp mills. Now he wanted to use the waste from cedar shake mills for this new OSB product that was competing with plywood in the residential construction market.

As Williston recounts in later correspondence, his efforts at the time weren't exactly met with open arms.

"A mill site on the lower Fraser River was secured and detailed plans were prepared for the construction of an oriented strand board plant. A construction firm, experienced in such work, reviewed the plans and was prepared to bid. At this stage the government said that their advice was that if I were to succeed, the plywood industry would be negatively affected and

David Ainsworth saw potential in the Cariboo's mixed stands of aspen and lodgepole pine.

that I was to stop all planning since no financial assistance would be available. Not too long after an American firm was given financial assistance to construct an oriented strand board plant at Dawson Creek. In all this time the only operator who monitored my progress was Dave Ainsworth."

In retrospect, Williston's plan for a coastal OSB mill had some timber supply challenges, but he was on the right track when it came to making full use of B.C.'s forest resources.

David, meanwhile, was also keeping tabs on OSB's progress in Alberta. Al Owen's company, Pelican Spruce Mills, built its first OSB plant in Edson in 1982. Owen quickly saw OSB's potential in the marketplace and had a second plant up and running in nearby Drayton Valley by 1985. Although there were other OSB producers by this time, their facilities were originally designed to manufacture waferboard and had been adapted to produce oriented strand board.

The Drayton Valley mill was Owen's pride and joy and he didn't mind showing people around.

"There was a lot of interest from others," says Owen. "It was a showplace. I was a hands-on guy and I made my tours short. I thought the more people who knew what happened here and could be introduced to OSB, the more mills would be built and the more we could sell our product."

Making full use of the company's aircraft, David made a number of trips to Drayton Valley in the mid-1980s, not only to educate himself about the manufacturing process, but also to familiarize others with OSB.

Whether he was accompanied by Steve Silveira, Doug White, Al Smith or various family members, David soon became convinced that OSB was worth exploring for the company's operations back in B.C.

In his view, acquiring timber for Ainsworth's sawmills was only going to get more difficult, which put their long-term viability into question. He also saw a time when the supply of large-diameter logs necessary for plywood would tighten up, making OSB that much more attractive as a substitute for plywood.

OSB now accounts for over 80 per cent of roof and wall systems in residential building in North America.

THE ADVANTAGES OF ORIENTED STRAND BOARD

The inventor of oriented strand board, or OSB, was Armin Elmendorf, who came up with the idea as far back as 1949 but didn't patent his invention until 1965. A product called waferboard appeared on the market in the late 1970s. It consisted of a random array of thin flakes about two inches square. OSB evolved from the waferboard industry soon after.

OSB is made from thin wooden strands arranged in layers. The core layers are aligned or oriented at right angles to the surface layers. These layers of cross-oriented strands are then bonded under heat and pressure with a water-resistant adhesive. Because of the oriented layers, this engineered panel has a structural integrity that is equivalent to plywood.

OSB is less expensive than plywood because it can be produced from low-cost pulpwood and under-utilized tree species, whether deciduous or coniferous. Also, OSB mills typically produce two to three times more product than plywood plants — using less than half the manpower.

OSB is designed for structural applications such as wall and roof sheathing, subflooring and webstock for the manufacture of I-joists. Oriented strand lumber, a thicker version of the panel, is now also being used in the construction of engineered homes.

Catherine, Brian, Allen and David Ainsworth in 1992.

The fact that OSB could be made from residue wood was yet another point in its favour; OSB would allow the company to use a whole new class of timber supply that wasn't suitable for lumber or plywood production.

Despite David's enthusiasm, he had an uphill battle convincing those around him that OSB was worth pursuing. The product was relatively new, and anyone getting into it would need substantial timber resources and up to $100 million in capital, given the massive plant required to make production cost-effective. It was also around the time of the Evans acquisitions, and company personnel had more than enough work to keep them occupied.

As Ainsworth's woodlands manager at the time, it would be up to Steve Silveira to tackle the problem of a wood supply.

"David was fascinated with the process of OSB and was saying that we've got a lot of small pine and aspen around 100 Mile that nobody's using, which could all go into making OSB," says Silveira. "I think all the rest of us had our doubts because we looked out there and saw poor-quality lodgepole pine and not a lot of aspen. But David was ready to go. He said he wanted me to start working on getting a timber supply for an OSB plant. So we launched very quickly into pursuing timber."

The earliest projections for timber requirements totalled over 500,000 cubic metres of wood annually — roughly 150,000 truck loads — which far surpassed anything available in Ainsworth's three Timber Supply Areas.

The company's initial conversations with forestry officials weren't encouraging. Chief among ministry concerns was the possibility of upsetting the other major licensees in the area. They would be opposed to any additional harvesting that might cut into their supply of merchantable timber.

Ainsworth eventually found an important ally in the regional manager of the Ministry of Forests, John Szauer, who was based in Williams Lake.

Szauer saw the potential of an OSB plant and suggested that the company pursue a long-term tenure arrangement known as a Pulpwood Area Harvesting Agreement. There were only a few such agreements in place in the province, and none had actually been used; they were created largely as a backup in case pulp mills ran short of chips. A Pulpwood Agreement tied to the construction of an OSB plant was a first for B.C.

Given the huge cutting area required, Ainsworth had an analysis done on problem forest types — wood that was below sawlog utilization standards but still occupied significant areas on the ground. The analysis also included deciduous trees that no one was using. This wood wasn't even included in the area's Annual Allowable Cut volumes.

The company's greatest challenge stemmed from the fact that this OSB plant would be unlike any other: it would run on a mix of lodgepole pine and deciduous trees. In fact, the species mix they envisioned at the time was roughly 70 per cent pine and only 30 per cent aspen. Al Owen's mills in Alberta ran on a diet of 100 per cent aspen, much the same as every other OSB mill on the continent.

Three decades earlier, David Ainsworth had pioneered the use of lodgepole pine, and now he was at it again.

"We knew early on that we had more lodgepole pine than aspen, and this was totally an aspen industry," says Steve Silveira. "This species mix wasn't what people were using. But David thought it could work. It was pretty much his vision all the way through; his vision and persistence kept it alive."

Regardless of timber availability, Ainsworth had to prove that they could manufacture OSB with a combination of pine and aspen and still obtain the necessary structural properties.

Al Owen, for one, was doubtful that the mix would work.

"I didn't think it was a good decision at that time for where he was because he'd be using pine," says Owen. "I had all my sights set on structural properties, and you don't want to get in there and produce something that's not a good structural product."

Even the residue from recently logged stands of lodgepole pine could be used for OSB.

David respected Owen's opinion and experience, but it didn't deter him from his goal.

"Al tried to tell me that it was not something we should try to get into," says David. "But by this time we were too far along, and we don't back out very easy — even with the advice of someone we greatly admire."

Keen to test their species mix, Ainsworth sent some 3,500 kilograms of area logs to a plant in Germany where they were processed and made into OSB panels. Combinations ranging from 70 per cent pine and 30 per cent deciduous right to the opposite were tried out, and subsequent testing showed the boards met all the necessary standards.

Ainsworth now knew they had a viable product.

Given the test results, Silveira and the woodlands staff estimated they would need a Pulpwood Agreement that could provide roughly 70 per cent pine, with a 30 per cent aspen component. The company felt it had done its homework — several years' worth actually — and regional ministry staff were now largely on board with the details of the proposal.

None of that seemed to matter, however, when officials in Victoria announced a drastic reduction in the volume of timber that they would make available within the Pulpwood Agreement. Ainsworth was seeking some 500,000 cubic metres annually. It was a significant amount of timber by anyone's standards, but all of it was from grades or species that were not being used, over an area that encompassed some 6.6 million acres and contained well over 30 million cubic metres of pulpwood in total.

Ministry officials said they would make only 330,000 cubic metres of timber available annually. And it would be on a backup basis only. The government's reduced offer was a far

---

I was in the lumber business before OSB and was going to build a plywood plant. When I got looking into it, I liked an idea I saw at a conference at Pullman University in Spokane. They were doing some stranding with a little six-inch hand machine. I thought if we could put that through a plant similar to one used for waferboard, and orient the strands, we could get a good structural product. So we fooled around with that in a little panel and worked our way up to three-foot-by-three-foot panels. Then I went over to Schenck in Germany and they had a little mobile plant there they used for waferboard, and we decided to try some orientators in that and run it through the press.

When I started OSB in Alberta they were taking those vast areas and turning them into pasture by mowing the aspen down and burning it. I could see all this aspen being wasted and thought we could use that to make some good product. And that we did.

My plants produced OSB from the very first board — the first ones to be certified in North America. Other plants were still running waferboard.

I always ran an open plant. As soon as I got my APA [Engineered Wood Association] approval they all came to see what in the hell us little farmers were doing up there.

We were pretty serious, working seven days a week. I did half the engineering right on site. I didn't have any consultants to rely on except the boys that helped me from Schenck, but they didn't have anything to do with our construction. They didn't know our building codes or procedures. I was designing as we went. I had my own concrete plant on site. I had nowhere to steal staff from and no extra money to pay anyone extra. I had to fight from the word go. And one plant was built in a slough.

I don't think I sold any board over $100 on a three-eighths-inch basis. Then the year after I sold out to Weyerhaeuser, OSB started to skyrocket and the price more than doubled.

— AL OWEN

Doug White headed up Ainsworth's woodlands staff as OSB gained increasing prominence.

cry from the 500,000 cubic metres David was anticipating, and he considered it a major blow. He even questioned whether the project was viable.

"They said, 'You don't really need all that,'" says David. "They trimmed us down considerably and we had to stand on our own two legs and say, 'No, we won't take that.'"

Despite the company's protestations, the ministry held firm.

"David did not want to take a 'no' to the 500,000 cubic metres," says Steve Silveira. "I was sent back and we pushed, but at the end of the day that was it. Take it or leave it."

The ministry also said the Pulpwood Agreement would have to be put up for competitive bidding. After all, the timber in question was a public resource. Although Ainsworth understood the need for a public bidding process, it meant even more time and resources would have to be expended in making a formal presentation.

There were other stipulations, as well. Before Ainsworth could harvest any of the wood in the Pulpwood Agreement area, they had to use up the small-size pulpwood in their existing forest licences. David didn't consider this a problem. He had already been looking at debris piles in the company's operating area and recognizing that there was usable wood being left behind.

The company would also be required to purchase any pulpwood that was available from other sources, at rates considered reasonable by the government. This too was something they had anticipated, given the fact that such clauses were tied to previous Pulpwood Agreements.

When the forest ministry announced in early 1988 that it was accepting bids for Pulpwood Agreement No. 16 (PA 16), the Ainsworths knew they would face stiff opposition from the major licensees in the region. The area in question overlapped the Timber Supply Areas of Williams Lake, Lillooet and Kamloops — the operating areas of at least four other major forest companies. And now everyone involved would get to express their views at a public hearing.

With a battle looming, David knew that community support would be a crucial element in their efforts to win the bid. A packed banquet room at the Red Coach Inn on a Saturday night in April 1988 was testament to the 30-plus years of goodwill that Ainsworth Lumber had generated in the South Cariboo region.

The company had called the meeting to lay out its plans for an OSB plant and seek the community's support in its upcoming bid for the timber.

As David outlined in his presentation, "We'll be needing the support of everyone, and it's better to have our friends and neighbours know what we're about so we can get that support if it's needed."

The company came prepared.

In addition to bringing the community up to speed on their bid for PA 16, they explained the OSB production process and showed off test panels just returned from Germany, as well as drawings of the proposed plant. The facility would cover 240,000 square feet and be equipped with the latest in anti-pollution technology. As with its studs, Ainsworth would focus on premium-quality OSB panels, even if it meant spending more capital. The audience also learned that negotiations were under way to purchase from Bridge Creek Estate a 40-acre building site immediately west of the Exeter sawmill.

David outlined the numerous hurdles the company still faced, not the least of which was financing such an endeavour, as well as penetrating the market for OSB. At this stage in the project, they believed their best prospects lay in selling premium-quality panels in Japan, deeming the North American commodity market too competitive. To this end, the plant would be the first on the continent to produce panels sized for both the North American and Japanese markets.

The presentation also emphasized the fact that the Interior lumber industry was in for hard times because of significant decreases in the amount of timber they'd be able to harvest each year.

"There are going to be tough times ahead, make no mistake about it," David told the audience. "We have been driven — and I don't use that word lightly — to use the part of the forest that is remaining."

Steve Silveira said the company's Annual Allowable Cut was dropping from 860,000 cubic metres to 650,000 cubic metres.

"Nobody likes to think about it, but it's a fact," Silveira told the group. "We need to find some kind of timber to replace the shrinking number of sawlogs. Within two years we'll be facing a 25 to 30 per cent overall shortage of sawlogs for the Chasm and 100 Mile plants. If we make OSB, we can use trees as small as three inches at the base and two inches at the top, greatly increasing the range of timber that is useable."

Various community leaders, from area mayors to the local Member of the Legislature, took the podium to voice their full support for Ainsworth, as did the president of the Cariboo Woodworkers Association.

David cited the company's pioneering work over the years and concluded that "this is going to be a very large project for us, but if anyone can do it, we can."

Nearly 11 months after Ainsworth announced its plans for an OSB plant in 100 Mile House and submitted its bid for timber, the public hearing for the awarding of PA 16 got under way in February 1989 at the town's community hall. It was a standing-room-only event as 300 spectators and participants converged to hear two full days of submissions and cross-examination from 80 sources.

The company was represented at the PA-16 hearings by (from left) Kelly McCloskey, Steve Silveira and David Ainsworth.

Ainsworth's bid faced stiff opposition from other forest licensees in the region.

With Ainsworth being one of the biggest employers in the area, there was plenty at stake for the company and community combined. Without the timber from PA 16, there would be no OSB plant locally, and the company's lumber operations were already feeling the effects of a reduced Annual Allowable Cut. David's loyalty to 100 Mile House was no small factor in the equation. He was deeply committed to seeing this plant built as a boost to the local economy, as well as a legacy of his four decades in the area.

Sitting in the hearing's hot seat representing Ainsworth were David, Steve Silveira and Kelly McCloskey. McCloskey had recently joined the company as chief forester following a stint as assistant manager of the Cariboo Lumber Manufacturers' Association. He was a Registered Professional Forester with a thorough knowledge of the industry, and brought considerable expertise to the team.

The hearing was chaired by Wes Cheston, B.C.'s Assistant Deputy Minister of Forests. There were only two bids for PA 16: Ainsworth's and one from Georgia Pacific, which was looking for 100,000 cubic metres of fibre annually for its chip mill at Cache Creek.

As expected, the bulk of the opposition to Ainsworth's bid came from other major licensees in the region. Their biggest fear was that Ainsworth would now have access to merchantable timber already in their supply area. They also opposed the fact that Ainsworth would have the rights to so much pulpwood or stands of timber currently considered uneconomical. Ainsworth, they claimed, would be tying up "the last available timber in the area."

As David and his team pointed out time and again, PA 16 would provide access to the pulpwood, but everyone had an opportunity to bid on it.

One major licensee argued that PA 16 would encroach on the licensee's already existing Pulpwood Agreement. The licensee also acknowledged that they had yet to utilize any timber in that Pulpwood Agreement area. Various plywood producers in the region, meanwhile, claimed that the OSB plant's output could lead to closures at their plywood plants. Other opponents said that Ainsworth's purchases of private wood would drive prices up to the point where their mills wouldn't survive.

The opposition to Ainsworth's bid wasn't limited to forestry companies. Area resorts and trappers pointed to the possibility of increased clearcuts, which could have an impact on wildlife habitat and tourism values.

The panel also received a wide range of submissions supporting Ainsworth's bid for PA 16 from area native bands, community leaders, businesspeople and politicians. Virtually all of them spoke of Ainsworth's long-time presence in the community and its reputation as a trustworthy operator and model corporate citizen for over three decades.

The company's proposal also detailed numerous benefits, including 176 new jobs created directly; over 350 additional new jobs created indirectly; a $60-million capital expenditure for the plant; $110 million in annual economic activity; income for farmers and ranchers who could sell timber to Ainsworth that would normally be cleared or burned; increased tax revenues for government; and reduced slash burning with better utilization of logging residue.

The company got the last word at the hearings, and David Ainsworth's closing comments played no small part in the eventual outcome.

"If I may, I will finish where I started off, talking about the process we went through in 1974 to secure the fibre for our sawmill at Chasm," concluded David. "I actually feel a bit of déjà vu. There we were, proposing to head off along a track of utilizing timber stands that, at the time, were not being harvested. Here we are again, proposing to construct a plant that will utilize timber stands currently in the same category. We are doing it again, in many ways, with the same number of uncertainties, unknowns and potential pitfalls. But like last time, our intent is sincere. It is our neck that is on the line and no matter what the future holds, if awarded the licence, we will make every effort to follow through on our commitment. Thank you very much for this opportunity to speak."

Despite the fact that the Ainsworth team was well prepared and presented a solid proposal, it was a tough two days for everyone involved.

The family business was about to undergo a major transformation as the company pursued expansion into OSB. From left: Allen, Catherine, Brian, Susan and David.

"We had tremendous opposition from the rest of the licensees at the hearing," says Steve Silveira. "They were saying, 'This will really cut into the future of the sawmill industry ... it's going to cause closures ... we won't have enough timber in the future because we're all counting on this.'

"Our counter was, 'Look, this wood is not part of the Annual Allowable Cut. Nobody's been using it and everybody had a right to put in a bid. If you really wanted this timber you could have bid on it ... we're sitting in this open process and we're the only bidder.'"

For his part, David was personally taken aback by the level of opposition from industry players and others. He felt that some of the arguments amounted to attacks on his integrity and implied that the company had other plans for the timber or would not live up to the stipulations set out by the agreement.

Several weeks after the hearing, Kelly McCloskey received phone calls from various government employees. They told him that they had never been through a public process where the community displayed so much support for the developer.

"Some of the government guys described it as a love-in," says McCloskey. "Certainly our staff didn't see it that way. It was brutal and long and the planning was exhaustive. But the love-in part was because so much of David was in front, and the community had so much time for him. There was so much goodwill that he brought to the table. His final speech came across really well and it clinched the deal."

Within two months, the government announced that Ainsworth would be awarded cutting rights under PA 16.

Now the real work began.

Ainsworth Lumber marked its 40th anniversary in 1992 with a day of festivities in 100 Mile House for all of its employees. Long-time service awards were received by employees from all five divisions.

David Ainsworth was front and centre in a 100 Mile House parade as part of the anniversary party.

The 330,000 cubic metres of timber a year granted to Ainsworth was far less than they asked for, but the decision was made to carry on. The next hurdle would be financing their venture. Before any potential lenders or investment partners could be approached, the company had to prove there was a market for their product. In this sense, the timing couldn't have been better. David Ainsworth and others in the company had pegged Japan as a key market from day one. For this reason, the 100 Mile plant was designed with a nine-foot-wide production line, which allowed for the three-foot-wide JAS-grade panels required by Japanese customers.

100 Mile's log yard crew aboard the Wagner loader in the mid-1990s.

As it turned out, Michael Ainsworth had just completed a year of employment with the Council of Forest Industries and was about to return to university for his master's degree. David had approached Michael earlier and suggested he look into the Japanese markets as part of his studies.

Professor Dave Cohen was new to UBC and had barely unpacked his books when Michael introduced himself and announced that he already had a research topic in mind.

"Dave Cohen was just trying to get his bearings and I don't think he was ready for a graduate student," says Michael. "Usually a professor has to settle in for a year or two before taking on graduate students and developing research topics. But I said, 'I have this topic already: marketing OSB in Japan.'"

With a company sales office in Tokyo and considerable contact with Japan already in place, Michael's thesis topic was a natural fit.

Beyond the university environment, however, the picture couldn't have been more bleak.

B.C.'s forest industry was in the throes of a prolonged downturn after recovering from the recession of the early 1980s. Sawmills throughout the province lost a combined total of $300 million in 1990, and it was late 1991 before conditions began to turn around. Ainsworth was hit hard over those 24 months like everyone else, and the OSB project slowed to a crawl as a result.

Soft lumber markets and a resumption of U.S. countervailing duties played a significant role. By 1993 the company had spent over $3.6 million on U.S. countervailing duties of up to 14.5 per cent on shipments of softwood lumber.

On the political front, the Social Credit government of Bill Vander Zalm was replaced by Mike Harcourt and the NDP. Environmental groups were putting forest industry activities under the microscope like never before.

With mounting public concern about the state of the province's forests, the government responded with plans for reduced cuts and more protected areas.

High stumpage rates were also taking a toll, particularly on Ainsworth.

In late 1991 the stumpage rates rose 32 per cent across the industry, while Ainsworth's rates increased by an alarming 237 per cent over several quarters. The company's fees jumped from $330,000 one month to a staggering $800,000 the next. The reason for the huge increase was that the province's complex stumpage system works on averages. Ainsworth's product mix of studs and pulp chips for export made it more susceptible to major swings in stumpage rates. The company had notified the government of this anomaly earlier in the year, expressing concern that without some kind of relief they couldn't continue to operate. The change in government and the resumption of U.S. duties, however, delayed a response.

Earlier in the year, to cut costs and raise some capital, Ainsworth sold its sales office building in Vancouver and moved Rod Lee and Blair Magnuson, plus two assistants, to a temporary sales office at the Abbotsford finger-joining plant.

By late December, the situation was so tight that Ainsworth was forced to temporarily lay off 150 logging and trucking contractors in the 100 Mile area until market conditions improved.

It was around the time that markets finally picked up that Michael Ainsworth completed his master's degree on oriented strand board in Japan. There had been minimal movement on the new mill during his time at school, and the question of how the OSB project would be financed was quickly coming into focus. David was temporarily sidelined by ill health and Michael was looking for his next step within the company, now on a full-time basis.

Ainsworth's management structure was as lean as it could be after the two-year downturn the company had just weathered. With no Vancouver office for the time being, Allen's Vancouver home became his base of operations and the centre of the OSB activity once Michael joined the project.

Michael was in charge of putting together a business plan for the OSB mill, armed with binders full of financial forecasts and budgets, proposed capital expenditures, market forecasts and profit-and-loss statements.

He and Allen then spent much of 1992 visiting every possible equity partner and lending institution that would listen to their proposal.

Although the markets had risen considerably, the major banks had no interest in participating in a $100-million project with a company the size of Ainsworth. Potential equity partners were particularly concerned about how Ainsworth's OSB would be marketed. As a result, Allen and Michael attempted to strike up partnerships with various existing OSB producers to market Ainsworth's product. Although they came close to forging a joint marketing plan, a potential partner in the U.S. backed out at the last minute.

Meanwhile, the amount of capital needed was well beyond the maximum lending limits

> *When we sold the office building near 1st and Burrard, the deal with the buyer was that we would be out immediately. This was a Wednesday morning when Allen came in and told us he had finalized the deal. We had been in that location about three years. So we called the Abbotsford finger-joining mill to prepare a space for a sales office for us out there. They took down a few walls in the office, and we called the telephone company with a special request and they put in phones for us. We called the afternoon crew at Abbotsford and they came in and we loaded up a rented truck. We moved out all our desks and files, all our records and personal stuff. By about 11 o'clock that night we were having beer and pizza and we went home.*
>
> *The crew went back to Abbotsford and hauled everything into the office that night, and Thursday morning Blair and I showed up for work at Abbotsford and it was business as usual. Well, not quite as usual.*
>
> — Rod Lee

Michael and Catherine Ainsworth were quickly growing accustomed to high-level financial meetings.

of the Federal Business Development Bank (formerly known as the Industrial Development Bank), which had always served Ainsworth well in the past. Every other major forest company that was approached declined the offer to participate.

"Allen and I spent a year searching for a partner. I don't know who we didn't talk to," says Michael.

"We had exhausted all of our possibilities," says Allen.

By the time their list of potential partners ran out, Michael and Allen had been turned down 88 times.

"We laughed about it all later," says Michael, "because they actually did us a favour. It became evident that to get a bank or somebody to finance or to do the debt side of this project, we were going to have to have some equity to throw at it. That's when the idea evolved that we were going to have to look at going public."

David and Susan had a long-standing relationship with Kamloops lawyer David Rogers, who had suggested in the past that it might be time for Ainsworth to go public — that is, to hold an initial public offering or IPO, sell shares to investors and become a publicly traded company on the Toronto Stock Exchange, complete with a board of directors and all the additional regulatory requirements that prevail.

For an independent-minded family accustomed to running operations their own way, going public wasn't at the top of anyone's list of preferences. But like it or not, they realized it was the only way to finance the move into OSB.

Rogers referred the family to Morley Koffman, who had a wealth of experience in taking companies public. Michael and Allen then embarked on the task of lining up underwriters. They eventually settled on RBC Dominion Securities, Burns Fry and Gordon Capital.

Satisfying the various requirements involved meant that going public and offering an IPO could take anywhere from 6 to 12 months. Fortunately, Michael's previous work on a business plan could now form the basis of a new prospectus designed to answer the questions of potential shareholders.

The Ainsworth "road show" to attract investors was a family affair from start to finish.

Site preparation and road building were the first order of business in the 100 Mile OSB project.

Allen would introduce the proposal, Michael would give an overview of the product and the markets, as well as financial details, and David would wrap it up. The entire presentation would last 30 to 40 minutes, and be followed by a question-and-answer session.

The company's initial goal was considered modest by IPO standards. It was only after the target was increased to $55 million that investors began to show interest.

The bulk of the IPO presentations were made over the course of a week in the spring of 1993. There were stops in Montreal, Toronto, Winnipeg, Calgary and Vancouver, with audiences in hotel meeting rooms ranging from 20 to 70 people.

Despite the numerous delays that had beset Ainsworth's plans for an OSB mill in 100 Mile — receiving the environmental permit took 18 months alone — the timing of the IPO was fortuitous. The markets for lumber, specialty plywood, veneer and chips had never been higher, and investors clearly saw potential in the forest sector. Ainsworth's IPO was for 5.5 million common shares. At a book value of $10 per share, the company successfully raised $55 million for the OSB plant.

Most of the shares were sold by the underwriters to institutional investors during the road show. By the time of the public offering on May 20, 1993, the company could have sold three times the shares that were available.

"The market was ripe for it," says Allen Ainsworth. "The shares started at $10 then soon went to $12 and they were up to $18 before another 24 months. There was a lot of support in the marketplace."

> *Dave was always a visionary. He was always trying to look ahead. The move to OSB tickled his fancy and turned out to be a very progressive move. The reaction in the industry at the time was, "Gee, this is a major step that is going to require an awful lot of capital. Is the company financed well enough to do it?" At that point I think I was approached by a number of the financial people that they had talked to, asking if the company could deliver on this. So we were all very positive about their abilities. They were always able to build and operate very successful plants.*
>
> *It was an encouragement to the financial community to support that business. And consequently they found their capital, their debt requirements for the business.*
>
> — CONRAD PINETTE

Although nobody knew it back in 1986 when Ainsworth purchased the Evans mills in Savona and Lillooet, those assets also played a role in the successful bid to go public. The two new divisions had pushed Ainsworth's annual sales over the $100-million mark. Attempting to go public with anything less than that would have been a hard sell.

As a result of going public, the company now had a board of directors made up of Allen as president and chief operating officer, Brian as chairman and chief executive officer, Catherine as chief financial officer and secretary, and David and Susan as directors. The three non-family directors were Morley Koffman, Gordon Lancaster and Gordon Green.

By owning a majority of the shares, the family retained control of the company and remained firmly in the driver's seat. But the decision-making process was now more formalized; going public meant adhering to all the rules and reporting requirements that come with being traded on the stock exchange.

"I was a bit torn on the issue," says Michael Ainsworth. "On the one hand I wanted to see us execute that project, and on the other hand, I was very conscious that we wanted to make sure we kept enough of the company so we could call the shots day to day. But we recognized that we would have to do some things differently, and maybe that wasn't a bad thing. It was going to bring a bit of structure and more accountability within the company, which probably was a good thing, especially when we were growing the company so aggressively."

With a successful initial public offering having been completed, site preparation for the 100 Mile plant began almost immediately. While visiting Al Owen's operations in Alberta, David had been impressed with engineering consultant Dave Wright, who had now been involved with the construction of six OSB plants. Wright was subsequently hired as project manager for the new Ainsworth mill, and Vancouver's Ledcor Industries was chosen as the construction manager.

Within a week the 40-acre parcel west of the Exeter mill was crawling with heavy

Sales team of Dawn-Ann Byers, Blair Magnuson and Rod Lee at their temporary Abbotsford office.

equipment brought in to clear the site, build roads and a log yard, install the underground services and begin work on the building's footings.

With the construction activity already in high gear just a few hundred yards away, David and Kevin Ainsworth represented the family at the official sod turning on June 7, 1993. Despite overcast skies threatening rain, more than 100 onlookers turned out for the occasion. David told the audience that "the plan all along was to build the very best panel board plant possible today. We've done our homework. And as I reminded one of my grandsons recently, don't let anyone, anywhere, tell you we don't know what we're doing."

David Zirnhelt, the district's MLA and Minister for Economic Development, said, "I'll bet it's not often that the barons of Bay Street have three generations of the same family saying, 'Have I got a deal for you.'"

Within several months of the sod turning, workers were pouring concrete for the plant's massive press pit. They dubbed it "the big pour." Crews worked 15 hours non-stop pouring over 1,150 cubic yards of concrete into the forms. The pit's foundation walls descended 25 feet into the ground and were 3 feet thick. Over 100,000 pounds of rebar were required for the job. The timing of the concrete deliveries was so crucial that special arrangements were made with BC Rail for that day to ensure that the trucks wouldn't be held up by train traffic at the Exeter crossing — one of the time-honoured features of driving out to the Exeter site.

Having spearheaded the company's OSB business plan, Michael Ainsworth was afforded the opportunity to learn about oriented strand board mills from the ground up. He began working closely with project manager Dave Wright and other engineers, visiting various plants and learning about the technical merits of one type of equipment over another.

As the family's point man on the project, he participated in negotiations with equipment vendors and the planning of the mill itself, as well as hiring some of the key staff members to start up operations.

David Ainsworth told the audience at the sod turning that "We've done our homework." Looking on were Mayor Ray Carlson, MLA David Zirnhelt and Kevin Ainsworth.

Ainsworth Lumber was about to enter the world of OSB.

Kevin Ainsworth at the official sod turning.

First among those to sign on were Lee Armstrong as general manager, John Barford as production and maintenance manager, and Jim Miller as process and technical control manager. Tyler Harris, who eventually rose up through the managerial ranks, was also part of this group. All four men had started with Al Owen in Edson, Alberta, up to 10 years earlier, although his two plants were now owned by Weyerhaeuser.

Beyond the managerial ranks, Ainsworth's newly expanded human resources team had its hands full when it came time to sign up employees for the new facility. Headed up by industry veteran Jim Butterfield, a new recruit himself who had just migrated up Exeter Road from the 100 Mile Weldwood operation, the team fielded a total of 1,600 applications for 100 positions. They even had an applicant from England. Not wanting to deplete their other

> Lee Armstrong approached me and asked if I'd consider working for Ainsworth Lumber. I was working in Edson and didn't know a lot about Ainsworth Lumber. He had plane tickets for my wife and me from Edmonton to Kamloops and had a car rental to go to 100 Mile and have a look at the project, so we decided to do that.
>
> I met Michael Ainsworth and Dave Wright that weekend. The project was still about nine months away from startup. When I came to 100 Mile I had no intentions of coming to work for Ainsworth Lumber. But when I left after the weekend, there was just something about the organization and the operation that I thought I wanted to work for them. They seemed like a good company to work for, an exciting company to work for, a company that was on the move and just getting into engineered wood products. All the people I met from Ainsworth and the family were very positive. They cared about the business, the people and their customers. I could see lots of opportunity both for myself and the company, so I made the move — and I'm very glad I did.
>
> After the mill started up, like most large start-ups we had a lot of call-ins after hours. Everyone involved worked very hard, and the owners were very patient. At one or two in the morning, when equipment would fail, people would always show up to help out with the breakdowns; there was never any issue. We worked a lot of hours the first six to eight months. I remember all of us working weekends for the longest time.
>
> — JOHN BARFORD

Large concrete pours had to be coordinated with BC Rail so trains wouldn't hold up the cement trucks en route to the construction site, which was adjacent to the Exeter sawmill.

divisions, they decided that no more than 50 per cent of the OSB positions would be filled by current employees. As it was, there were over 170 in-house applications; they could easily have filled the OSB roster without going outside the company.

With plant construction going full steam ahead, Ainsworth's forestry staff was well into the process of sourcing out timber for the mill. For starters, it meant wrestling with the nuts and bolts of how PA 16 would actually be implemented.

Kelly McCloskey had since left the company to represent Ainsworth and several other major licensees at regional land-use talks. His departure opened the door for Tim Ryan to join the company as its new chief forester. Like McCloskey before him, Ryan had also served with the Cariboo Lumber Manufacturers' Association before joining Ainsworth. Prior to that, he had been a forester with B.C. Forest Products.

His first assignment after joining Ainsworth in September 1992 was securing an extension for PA 16, which was nearing the end of its five-year term.

"On any of these agreements, you're given the rights to the timber under condition that you're going to build a facility in a certain period of time," says Ryan. "If you don't, you're normally given some opportunity for an extension. After that second chance, though, it gets more difficult to get an extension. The Crown and the public want to see an investment made because it creates jobs."

By this time the forests ministry bureaucracy was fully onside with Ainsworth's OSB proposal, from 100 Mile

> *When the first mat went into the press at 100 Mile, the press started shaking in ways it shouldn't shake, and the German reps started running. The press was doing some wonky things and they were getting the hell out of there in case we blew some lines or something. It was going into high pressure too early, so it kept jumping and banging, and the hydraulics and everything started shaking. There was a lot of tonnage and pressure involved. So once everybody saw the Germans running, everybody else started running. By this time it was in the wee hours of the morning. Nothing really went wrong with the press, but that was our first press load.*
>
> — JIM MILLER

Construction work at the plant's log infeed decks and log ponds.

District Manager Ray Ostby to Assistant Deputy Minister Wes Cheston in Victoria. And regardless of whether it was Social Credit or NDP, the government of the day was supportive.

"In the forests ministry there were some strong individuals with strong personalities who carried the day," says Tim Ryan. "When other licensees said the wood wasn't there or that the sky was going to fall, the ministry guys said no, there's a clear opportunity here to utilize this unutilized timber out there. They knew the PA area and the stands and that it wouldn't be an over-cut situation."

Ainsworth was duly awarded a two-year extension on PA 16, but that didn't mean the work was over. The company was also required to submit timber management plans for PA 16, part of which involved identifying the various stands that were located in the cutting areas of other licensees.

With five forest districts and two forest regions involved, PA 16 overlapped the operating areas of at least five major licensees, none of which supported the proposal from the beginning. If they now chose to make the planning process difficult, they had the means to do so.

A controversial land-use planning initiative called CORE (Commission on Resources and Environment) added another layer of complexity to the PA 16 proceedings.

The idea behind CORE was to bring together all the stakeholders in the Cariboo-Chilcotin region, from forest companies and local government to ranchers and resort owners, and come up with an overarching land-use strategy that was based on the principles of sustainability.

The completion of Ainsworth's first OSB plant in 100 Mile House marked a new era for the company.

The objective of CORE was to develop a land-use plan for the Cariboo-Chilcotin by consensus. It was an admirable goal but nearly impossible to attain considering the diverse cast of participants.

Ainsworth and other industry players were represented by Kelly McCloskey, who participated in over 60 days of meetings spread out over two years. Trying to sort out opposing views on numerous issues made the job of long-term timber management in the region more complicated for those in the forest industry.

The regional CORE participants never did reach consensus, so the commission staff submitted their own report to the government. The result was a disappointment to everyone involved and triggered another round of unofficial talks in an effort to create a "made in the Cariboo" solution.

Farther south in the Lillooet portion of the PA 16 area, one of the province's more contentious areas when it comes to land-use planning, Ainsworth's foresters were involved in a variety of negotiations and round tables concerning Protected Area Strategies and Old Growth Strategies. And many of the areas affected by these initiatives overlapped with areas of concern to the region's First Nations.

Ainsworth staffers Mark Hopkins, Don Brown and Steve Law did a significant amount of work on the implementation of PA 16 in that region.

Lining up the timber sources to feed this new OSB mill involved more than PA 16, however. Since the Pulpwood Agreement was strictly for backup purposes, the onus was still on Ainsworth to locate long-term sources of aspen and pine — even if the pine was of a grade no one else wanted. The South Cariboo's aspen stands weren't sufficient for the wood volume required, so Doug White, who now headed up Ainsworth's expanding woodlands department, had to look elsewhere.

> *The first day I was in 100 Mile doing my job, I was walking by this older gentleman — it turned out to be David Ainsworth — and he said, "Who are you?" I said, "I just started today. I'm the new process and technical manager." He said, "You are? All right, come with me." So he took me into the coffee room and he sat me down and he had about a 10-minute conversation with me. It was about Ainsworth's reputation and how good their reputation was in the forest products industry. He said they enjoyed having it and they wanted to have the same reputation with OSB — they wanted to make a good-quality OSB product all the time.*
>
> *It was an excellent conversation to have with the gentleman who founded the company. It sent me headed in the right direction right off the bat. It laid the groundwork for me that said, "Hey, you work for a company that's not going to nickel and dime resins or densities or anything like that."*
>
> *From that day forward I've told numerous people this story because it's what we do.*
>
> — JIM MILLER

Mike Leach and Jim Miller in 100 Mile's original control room.

"Watching this OSB plant being built was a huge thing," says White. "It was my responsibility to get wood for it. We knew there wasn't that much aspen around 100 Mile and there would have to be a fairly large purchase program. I spent a lot of time going around the Prince George area and convincing people to cut down some aspen trees and ship them to us."

It didn't take long for the idea to catch on, as landowners watched their neighbours earn extra dollars by clearing a few acres and shipping the aspen south to Ainsworth.

"The average cut early on might be just two loads," says White. "We'd have 400 to 500 suppliers in a month, which meant hundreds and hundreds of timber marks — lots of paperwork. Then sometimes you might get 500 loads from one guy."

As the purchase program caught on, it wasn't long before some 90 per cent of the plant's wood came from the Quesnel and Prince George areas.

Bringing in the low-quality pine, however, was another story.

"Other foresters were probably thinking we were all nuts," says White. "They were thinking we couldn't log all this small pine and haul it in there. I'm sure people thought it was going to be a disaster, a catastrophe. But I had faith in Dave. I had been to Alberta and knew you could grind up the aspen and make OSB. The hurdle was to get the conifer in there. And I don't think a lot of the skeptics really knew what an OSB plant was."

> *When we were making our first board in 100 Mile, we planned on having it in and out of the press before six o'clock in the evening, so Lee Armstrong had planned a big dinner at the Red Coach Inn to celebrate the first board. Of course, first board never happened at six, so he moved the reservations to nine, and it didn't happen at nine either. We were still trying to get the first board through at midnight, so we ordered pizza. We ended up making first board between four and five in the morning. So we all went for "first board breakfast" instead of dinner that night.*
>
> — JOHN BARFORD

It was Doug White's job to familiarize the new OSB employees with the area's wood supply and what they could expect over the coming months. Although the production staff knew that using a mix of species was part of the plan, White's mention of including birch in the production process caused more than a few raised eyebrows.

"I thought half the crew was going to quit; they were all looking at one another," says White, recounting his presentation. "It came as a real shock to them."

As the plant's new process and technical manager, Jim Miller would be one of those most affected. "We knew we were going to be using a lot of pine right up front, but Doug indicated we'd be using up to 10 per cent birch. That worried us a little bit for sure, because when we used birch in Edson it curled up just like a layer of a cigar."

As they later discovered, the mill's stranders and dryers were able to handle the birch without a problem; the key was feeding the birch into the mill at a consistent volume.

By the end of 1993 the plant's structure was fully erected and 85 per cent of the cladding was on. Some of the plant's components had already arrived from overseas and were sitting in storage, ready for installation. In fact, construction was going so smoothly that the startup was moved ahead one month to July 25, rather than August.

Following a month of testing and dry runs, the $105-million plant's main systems were commissioned and the first board was officially produced on August 24, 1994.

David Ainsworth was justifiably proud when he took the podium — fashioned from the mill's OSB — at the plant's grand opening on September 24, 1994. Over 1,200 guests turned out for the occasion, including Ainsworth employees from around B.C., politicians, equipment suppliers, industry representatives and residents of the area.

For David, it was a time of looking ahead, but also looking back.

A proud moment for the Ainsworths as David, Susan and Allen look on at the plant's grand opening.

The plant for him was the culmination of more than 40 years in the Cariboo, starting with a portable sawmill and a handful of employees.

He told the crowd he was especially appreciative of numerous long-time employees who had been with Ainsworth Lumber for three decades or more, pulling together as part of the extended Ainsworth "family" and contributing to the company's growth and success over many years.

Capping off the ceremony, David and Susan, along with 10-year-old grandson Steven Fellenz, cut the ribbon stretched across a plant entrance and invited everyone in for a look around.

Despite the lengthy delays along the way, starting up the new plant in 1994 turned out to be another case of fortunate timing for Ainsworth. The markets happened to be at a high point, and the company was able to make money within three months of startup, running at only 65 per cent capacity. Had the plant been completed much earlier, when OSB markets were weaker, it's safe to say that stress levels would have been significantly higher during the ramp-up phase.

Even the great quantities of unused lodgepole pine sitting in the log yard were less of a problem than would normally be the case. It was taking much longer than predicted to incorporate the pine into the plant's production mix, and storage space was running low.

David and Susan cut the ceremonial ribbon, with grandson Steven Fellenz looking on.

Brian Ainsworth, Derek Welbourn and Kevin Ainsworth.

Catherine and Michael Ainsworth.

The plant's grand opening drew hundreds of guests.

Fortunately, chip prices were also high at the time. Rather than clog up the log yard with inventory, the company was able to turn the pine into chips and sell them to pulp mills.

By this point the investment community was looking favourably on the company. Barely seven months after going public, Ainsworth leveraged its position to raise an additional $50 million U.S. through long-term debt placements, or bonds.

The bulk of their capital needs for the project were now covered.

Alberta's abundant stands of aspen made the province an attractive prospect for further investment.

If anyone wondered where Ainsworth's foray into OSB production was leading, they didn't have to wait long to find out. In fact, the wheels were already in motion for the company's next move before the 100 Mile plant had pressed a single panel.

The 100 Mile plant was still under construction in January 1994 when the Alberta government announced a request for proposals on 580,000 cubic metres of deciduous timber in the Grande Prairie area, about 250 miles northwest of Edmonton.

The Ainsworths already had their eye on expansion opportunities in Alberta. Not long before, they had made an offer to acquire a plywood plant in Edmonton, but the deal never materialized. In 1993 they came close to bidding on a deciduous timber allocation for an OSB plant in the High Prairie area, but ultimately decided against it. That proposal eventually went to Tolko, which built an OSB plant there in 1995.

When the Grande Prairie allocation was made public, the Ainsworths wasted no time in surveying the area's timber. Four months later, they came to a decision.

The company's chief forester, Tim Ryan, was among the first to find out.

"In April 1994 the Ainsworths came to 100 Mile, sat me down in the boardroom with Steve Silveira and said, 'Tim, we're interested in some timber in Alberta and Grande Prairie — go see what you can find out.' And away we went. We had until the end of May, just over five weeks, to put together a proposal and make people aware of who we were."

For a company that had recently gone public, was new to OSB and had barely completed turning the sod at its new 100 Mile plant, pursuing the Alberta bid was a bold move in anyone's book. The Ainsworth reputation wasn't built on shying away from opportunities, however, and they weren't about to start now.

The timeline to submit a bid was extremely tight — three months is the norm in such cases — but there were bigger concerns. For the first time in company history, Ainsworth was bidding on a resource that someone else actually wanted. As a relatively small B.C. company coming into Alberta, they were definitely the outsiders and were unknown to the Grande Prairie community.

The competition for the timber came from two U.S.-based leaders in the forest products industry: Weyerhaeuser and Louisiana-Pacific (LP). Of the two, Weyerhaeuser was by far the more formidable. They already had OSB mills in Edson and Drayton Valley, Alberta, and had been in Grande Prairie for nearly 25 years, enjoying strong corporate ties to the community. They operated a large pulp and paper mill and played a major role in Grande Prairie's economy. In addition, their foresters had several decades' worth of information on the area's wood supply. They, too, saw an opportunity for an OSB mill, as well as a small hardwood plant.

As the world's largest producer of OSB, Louisiana-Pacific certainly had the wherewithal for the project and was familiar with the area's timber, but its closest presence was a sawmill in Dawson Creek, B.C., just over the provincial border. LP also wanted to build an OSB mill with the aspen now being put up for bid.

As the undisputed underdog in the competition, Ainsworth had a lot of ground to cover in a very short time. And none of this was made any easier by the fact that only a minimum of forestry data was publicly available through government ministries.

First, they had to provide their own estimates of how much timber was available in the area. If they found more than the government was offering, they could propose more. If they found less, they could propose less.

Second, they had to detail the volume they would log each year and, most importantly, how much money they were willing to pay the Alberta government for it.

Third, and in addition to arranging the financing, they had to locate a site, then design and cost out an OSB plant, the size of which would be determined by the annual volume of timber available. Last, these newcomers from British Columbia had to make themselves known in the Grande Prairie area almost overnight.

> *I was the first person hired for Grande Prairie, and it was a process of two interviews. Your first was your standard interview with a panel of three, but the second interview was much more interesting. I had to go one-on-one with Tim [Ryan] for about two and a half hours. Allen Ainsworth was there too, and he kept popping in and out of the room. He had a dog with him, and he kept sending this dog into the room. I think what he really wanted was for Tim to end the interview so they could catch a plane. It was very unusual sitting there, and there's the president of the company coming in and saying hi, and then about half an hour later there's his dog coming in and bouncing around the interview while I'm trying to keep my focus. And if you know Tim, it's usually pretty intense discussions. That was one of my first experiences with the company.*
>
> *We had two floors of temporary office space downtown, and for about three months there was just me on the upstairs floor. I had a huge area, more like a huge meeting room. Just one desk in the corner and my phone and my pencil. That was it. I was getting hundreds of phone calls. They were either calling for a job, to sell us wood, or they wanted to be a contractor.*
>
> — DAVE COOK

It helped that Tim Ryan had a few contacts in the province, having graduated from the University of Alberta in Edmonton and worked for two years in Alberta before relocating to B.C. Preparing the timber side of the company's bid involved a hectic five weeks of back and forth from B.C. to Alberta, all the while attending various CORE land-use meetings as well.

"We'd spend our day working on B.C. CORE issues, then that afternoon or night we'd roar back to 100 Mile, jump on a plane and fly to Grande Prairie to attend a city council meeting and explain who Ainsworth is," says Ryan. "It was very compressed. We were given free rein if we needed to go to Alberta. We'd just phone up David and he would fly us over."

Ryan and his woodlands team made a point of meeting with as many environmental and citizens' groups as possible, as well as with First Nations. Not only did they learn about local land-use issues, but they were also able to answer questions about the company's plans for the timber and OSB production.

Back in Vancouver, Allen, Michael and Douglas Ainsworth handled the production and business end of the bid even as construction continued on the new plant in 100 Mile.

If awarded the timber rights, Ainsworth proposed to build a $135-million OSB plant that would produce over 540 million square feet of OSB annually. The plant would also include value-added facilities to produce engineered I-joists, finger-joined lumber, and joinery stock for windows, doors and mouldings.

The timber side of the bid, however, was the true indicator of how much Ainsworth valued the opportunity. In bidding for 510,000 cubic metres annually, the company proposed a cash bonus, paid up front, of $10 per cubic metre. A payment like this was unheard of in Alberta. In addition, they were offering another $6 cash bonus for every cubic metre of wood harvested off their Crown licence for the next 10 years.

The extra bonuses Ainsworth was including added up to over $20 million.

"We were saying, 'Yes, we're a new player here, but we're a quality player and we're going to pay money for this timber,'" says Tim Ryan.

As rich as the bonus may have seemed to Alberta observers, it was far less extravagant when viewed from a B.C. perspective. The Ainsworths were coming from an environment with high stumpage rates and U.S. countervailing duties. The B.C. timber supply was far more limited, and the costs reflected it.

Looking at Alberta's timber supply over the long term, the Ainsworths felt it would only become more valuable. Indeed, David Ainsworth had figured this out 30 years previously

Public information displays were part of Ainsworth's campaign to familiarize Grande Prairie residents with the company and its proposal. From left: Ian Hamilton, Tim Ryan and Mark Hopkins were all on hand for mall duty during the run-up to the bid.

Ainsworth's bid for 510,000 cubic metres of deciduous trees opened the door for expansion.

when lodgepole pine was still considered a weed species. Offering a premium price now to get in the game would eventually pay for itself many times when prices inevitably increased.

The bonuses on the timber weren't the only incentives Ainsworth used to swing the deal their way. Several initiatives in their woodlands program went above and beyond the government's requirements of the day. The first was the creation of an advisory body of local residents that would have input into harvesting plans. Although such panels were mandated in Alberta's Forest Management Agreements, Ainsworth was the first to establish one in this particular type of timber tenure agreement.

The company also committed to hiring a forester strictly to oversee private woodlot activity. Equally important was the establishment of an integrated resource management plan, which included funding for environmental and ecological research. This was a major initiative and unique to Ainsworth at the time.

Meeting the bid deadline by the end of May was a victory in itself, but the work was far from over. The run-up to the government's decision also included a public hearing in June, where the applicants presented their bids and fielded any questions.

Throughout the bid period and right up to the decision time, Ainsworth wasn't leaving anything to chance. Staffers continued their campaign to win the public's confidence with mall displays, an open house at the Grande Prairie Inn, ads in the media and meetings with a variety of groups, from environmentalists to native bands.

For their part, Weyerhaeuser Canada ran an aggressive bid campaign of their own, highlighting their ties to the community over many years, along with ready access to the necessary capital. They had a site, a project team ready to go, and they

Grande Prairie's daily newspaper announces the successful bid.

were committed to breaking ground the day after the bid announcement. Weyerhaeuser's proposal included a $174-million OSB plant and sawmill, while LP was offering a $90-million OSB plant with two I-joist plants and a panelling plant to follow.

The government's decision was announced to a packed meeting room on August 9 by the province's Minister of Economic Development and Tourism, Ken Kowalski. "The bonuses and the amounts provided to us by Ainsworth will give us the greatest amounts of return we've ever had for a deciduous product in the history of Alberta," said Kowalski, alluding to the fact that Ainsworth's bid topped their nearest competitor's by over $16 million.

In the end, Ainsworth was awarded cutting rights to 583,000 cubic metres of deciduous trees annually in return for a commitment to build a $129-million OSB plant. They had settled on a 160-acre site roughly 30 minutes south of Grande Prairie.

"Learning that we won the bid was almost overwhelming," says Allen Ainsworth. "There was so much pressure building up to the final decision. It was the biggest thing that had happened in northern Alberta in years, so it was featured in all the media."

Looking back over the months of preparation, Allen adds, "We certainly would have gotten an 'A' for effort if we hadn't won. But in fact we pulled it off."

The majority of local business people would probably have preferred to see Weyerhaeuser get the nod. They were good corporate citizens and well-established, while Ainsworth was an unknown entity with limited experience in OSB. Nevertheless, most people accepted the government's decision and welcomed this economic shot in the arm that the community of 30,000 sorely needed. The government's selection process was a model of transparency, and Ainsworth's bid was clearly deemed superior.

Ainsworth's arrival also meant more diversity in the regional economy — a key point for politicians — which was still recovering from a serious slump in the oil and gas industry. The plant would pump an estimated $50 million a year into the Grande Prairie region and provide over 800 jobs directly and indirectly.

> Gerry Bauer from Canfor said he had a site for our OSB plant south of Grande Prairie. They had looked at moving their sawmill there to get it out of town. The site was well positioned relative to our timber supply and away from people. Dave Wright came over that July with me, once we had chosen the site, and we went out there. He said it was a good site. One of the questions was do we have water? He says, "Well, I'm going to witch." So he went over to a willow tree, cut a piece of the tree off, trimmed the branches and started walking around witching for water.
>
> I said, "What the heck are you doing? I don't believe you." He looks at me like I'm a total disbeliever. So we walk around the site for a while and he's got his tree fork, and then the fork started really shaking. I said, "No way, you're bullshitting me!" They drilled right there and found water.
>
> — Tim Ryan

"The bonus bid really set the tone," says Tim Ryan. "It showed that Ainsworth was serious and was prepared to put their money where their proposal was. It must have been a very difficult political decision; I have to give them credit for that. But they chose us for the right reasons. When you stack up the proposals, ours was the best."

And though the bid had all the hallmarks of a David vs. Goliath scenario, "We never really viewed it as that," he adds. "We felt very confident in what we could put forward."

Ralph Klein's government, just a few years into its first mandate, wasn't about to give away the farm, though. To qualify for the timber, Ainsworth would have to start breaking ground within 30 days.

The company had anticipated the winner would have to begin construction reasonably soon, but 30 days after the announcement came as a shock.

Ainsworth also had to post a $5-million irrevocable letter of credit to ensure its project would be completed by the March 31, 1996, deadline.

Grande Prairie's press was manufactured in Germany. The press has 12 openings for master panels measuring 12 feet by 24 feet.

Left: Hot air from the plant's energy system flows through this duct to the strand dryers. Above: Constructing the storage bins where strands or flakes are held before entering the dryers below.

The plant, which is roughly nine acres under one roof, was constructed in only 13 months.

The 100 Mile OSB plant was now up and running, so Douglas Ainsworth, who would be the family's point man on the Alberta project, and contractor Dave Wright set up shop in Grande Prairie to oversee the construction of Ainsworth's second OSB plant.

The company didn't want to overplay its hand as a newcomer to the area, so Ainsworth's original bid proposal was actually on the conservative side, both in terms of the available timber and the size of the manufacturing plant. Now that they were more familiar with the area, Tim Ryan and his team felt there was an additional 200,000 cubic metres of wood available per year: over 150,000 cubic metres from private lands, and over 50,000 cubic metres from oil and gas salvage programs.

This increase in fibre supply, for a total of some 800,000 cubic metres annually, meant that Ainsworth could build a facility capable of producing over 540 million square feet of OSB per year — making it significantly larger than the 100 Mile operation.

By November 1994, Dave Cook was hired as Grande Prairie's woodlands manager, making him the first employee at the new division. Cook had been with the Alberta government for 13 years and was previously the chief forester in the Peace River District. Within a short time he had an 11-person staff signed on — all but one from the area or from northwest Alberta — and the woodlands work began in earnest. There were harvesting plans and permits to develop, and contractors had to be hired for logging and hauling, building and maintaining roads, and carrying out the necessary reclamation work.

As it turned out, Alberta was trimming its civil service at the time. This meant Dave Cook had his pick of senior woodlands people with a high level of expertise in government forestry policy as a complement to staff with an industry background. "We started off with a very good, professional and well-rounded group of individuals," says Cook. "It really helped our program succeed. Having that local experience, we knew the lay of the land. We were able to get going quickly because we knew the local forests, the access and the issues. We had that cross-section of experience, combined with the local experience, and it really helped us a lot."

When it came time to hire contractors, over 300 prospective candidates turned out for an information meeting that introduced the company and outlined its requirements in the woods.

"The attendance and interest exceeded our expectations," adds Cook. "There were operators there who owned one skidder along with companies that harvested over a million cubic metres a year."

Interested contractors were asked to submit a detailed business proposal, complete with financial statements. This way Ainsworth could select a combination of both large and small operators, giving the company a balanced configuration of equipment and resources that offered the greatest flexibility in changing conditions. By industry standards, it was a unique way to

Lunch time for Grande Prairie's startup crew.

Employees were on hand to see the first mats go into the press on December 14, 1995.

select a team of contractors. Many participants commented that the act of writing down a business plan was a useful exercise.

By deadline time, woodlands staffers had more than 115 submissions to pore over — a stack of documents that was 15 feet high.

On the plant's production side, the company's human resources team fielded some 5,000 applications for 165 positions. Special consideration was given to the region's native and Metis community by holding workshops on how to write resumes and prepare for interviews.

Allen Ainsworth holds the ribbon as Minhas Bros. Trucking brings in Grande Prairie's first load of logs.

Roy Bickle, our vice-president at the time, was very good at celebrating our milestones and successes as we were going forward, so we decided to celebrate the first load of logs coming into our yard. We made quite a do about it. We had the media and all the local politicians and one or two ministers from the Alberta government there when we cut the ribbon and brought the first loads over the scales.

It was pouring rain and everyone was in their rubber boots in the mud. Then it turned out that the first two loads that we brought in were grossly overweight because we couldn't properly gauge how much weight was on the trucks yet. We had a system set up on our scale house that, when you had a really large load, all these red lights would go off indicating that you had broken the law in terms of load weights. And both of these loads set the lights off like crazy. So somebody asked the question, "What are those red lights for?" and we said, "It's part of the celebration."

Then we had a big gathering afterward and we hired a local lady to make up a beautiful cake of a log truck. We had some members of the Ainsworth family there to cut the cake, but when I looked — we never saw the cake until the last minute — the spelling of Ainsworth was actually ANISWORTH. So we had to do a little bit of scraping there. All our dignitaries were there — it was pretty funny. The family thought so too.

— Dave Cook

By December 14, 1995, the first batch of master panels was rolling out of the press — barely 13 months after construction began. Project manager Dave Wright said it was probably the swiftest construction of an OSB plant in North America. At 415,000 square feet — roughly nine acres under one roof — the Grande Prairie facility was the largest single-line OSB plant in the world at that time, with approximately 50 located in North America alone.

David Ainsworth's vision, coupled with his determination to take the company into OSB production, had opened doors that few would have imagined possible just a few years before.

First with 100 Mile and now Grande Prairie, Ainsworth Lumber had entered the OSB industry in dramatic fashion.

The move into OSB came with a steep learning curve, but there were now three generations of Ainsworths, all based in Vancouver, sharing the load and developing a collective expertise that would soon compare with any in the industry.

In the meantime, going public hadn't dampened the family's entrepreneurial spirit, and with two new facilities in operation, they were ready to continue building the company.

The majority of logging activity takes place during the winter, when frozen conditions improve road access to wet areas.

Chapter 9

A NEW ERA

A NEW ERA

Another load of logs ready to be sorted at the Lillooet division log yard in the mid-1990s.

Ainsworth's emergence as an OSB producer was certainly front and centre during much of the 1990s, but its lumber, plywood and veneer divisions still played a vital role in the company's performance. In fact, it wasn't until 1998, when Grande Prairie was running at full production, that sales of OSB exceeded those of the company's solid wood divisions at Exeter, Chasm, Savona, Lillooet and Abbotsford.

As 100 Mile OSB was coming on stream in 1995, the sales of lumber, specialty plywood and wood chips still accounted for approximately 70 per cent of the company's annual revenues, with roughly $181 million in solid wood products compared to $77 million for OSB. By 1996, OSB had crept up to 40 per cent of total sales; by 1997 it was 42 per cent; and by 1998 it reached 57 per cent of sales.

The company produced 225 million board feet of lumber at Exeter, Chasm, Lillooet and Abbotsford in 1995. This included 60 million board feet of finger-joined lumber at Abbotsford. Savona's plywood output by 1995 was 70 million square feet.

The move into OSB, and the expansion into Alberta, looked particularly prescient when viewed against the backdrop of B.C.'s forest industry. The period around the mid-1990s was as tumultuous as any the sector had seen, and this was coming off a severe two-year slump.

The Exeter sawmill, shown in 1990, was producing over 115 million board feet of lumber at its peak.

Public concern over forest practices, particularly at the coast, was at an all-time high. A newly elected NDP government under Mike Harcourt responded with the ill-fated CORE process and, more importantly, the introduction of a new Forest Practices Code.

Faced with the spectre of intensifying boycotts in the U.S. and Europe, as well as protracted disputes on their home turf, many in the industry supported the new code. It was officially instituted in 1995 as part of the Forest Renewal Act. Within a short time, however, the true costs of this new legislation were becoming apparent.

The new Forest Practices Code was a maze of red tape and excessive bureaucracy. Relying on tough enforcement rules rather than offering incentives for good forest management, the code had an impact on every aspect of woodlands activity.

Forest companies bore the brunt of the costs. More staff had to be hired to navigate the code's complex guidelines, and consultants had to be hired to do the work previously carried out by staff. Ainsworth's woodlands department in B.C. grew from 35 employees in 1992 to 75 after the new forestry act came into effect. Out in the field, activities such as building roads doubled or tripled in cost, with an engineered bridge alone ringing in at well over $125,000.

The new forestry act didn't stop there, however.

As a response to the U.S. softwood lumber dispute, the provincial government created an agency called Forest Renewal B.C. (FRBC) as a means of returning tax revenues to the industry. Unfortunately for the licensees, FRBC was funded through stumpage increases, paid by the companies themselves.

Between new code-related costs and the increased stumpage fees, licensees such as Ainsworth were now paying an additional $35 a cubic metre for their wood.

The result was nothing short of devastating. With log costs among the highest in the world — up to $85 a cubic metre at one point — B.C.'s forest industry was growing increasingly uncompetitive in the global marketplace.

Few operations in the Interior felt the effects of these changes more than Ainsworth's sawmill and veneer division at Lillooet. The designation of new park land — 460,000 acres in the Stein Valley alone — and additional protected areas had already cut sharply into the company's available timber supply. The new Forest Practices Code and stumpage increases would add significantly to costs that were already among the highest in the industry. Submitting a Forest Development Plan to the government, for example, was 10 times more costly and time-consuming than just five years earlier.

In the wake of all these restrictions, more of Ainsworth's harvesting activity around Lillooet had to be shifted to steeper slopes, which meant greater reliance on high-lead cable logging and helicopter logging. In cable logging, the felled trees are attached to an overhead cable system powered by engines, which transports them down the slope to a central landing. It's a preferred method in sensitive areas where minimizing damage on the ground is important, or when steep slopes render skidder work too dangerous. Selective cable logging is also employed to preserve mountainside viewscapes.

Ainsworth's Chasm division sawmill underwent an $8-million upgrade in 1995.

Roughly 65 per cent of Ainsworth's harvesting activity in the Lillooet Timber Supply Area now had to be done with cable logging — a threefold increase over previous years. Because cable logging is more labour-intensive than conventional skidding and also requires more expertise and monitoring, Lillooet's log costs shot up even higher, reaching their current point of $100 a cubic metre.

Another aspect of Lillooet's Timber Supply Area involved the region's First Nations. Encouraged by recent Supreme Court decisions on native land rights, aboriginal groups were interested in having more input in land-use decisions, exercising more control over the land base and establishing joint ventures to participate more directly in the industry itself.

Kevin Ainsworth's duties had shifted from woodlands accounting and investor relations in 100 Mile to overseeing the B.C. woodlands activity. Along with environmental, Forest Practices Code and stumpage issues, he also became more involved in negotiations with First Nations throughout the company's operating areas, particularly in the Lillooet region.

"I had to learn to be patient and see other people's points of view," says Kevin.

"Whether it's negotiating access to an area, or working on a joint venture, the most important aspect of the process is building the relationships. After you've had a chance to attend the meetings and spend some time together, there's a deeper level of trust among everyone involved."

Despite a number of successful joint ventures still in operation, as well as other agreements with local First Nations communities, the question of Ainsworth's long-term timber supply remained in the forefront.

High-lead cable logging is used in sensitive areas near Lillooet.

By 1996 the company's log costs had risen by a staggering 52 per cent over the previous year. Despite strong markets and increased sales to Japan, producing lumber had become a money-losing proposition.

The Ainsworths' forecast several years earlier of a reduced log supply for the Exeter and Chasm sawmills was also proving accurate. On top of that, the U.S. implemented a new quota on imports of softwood lumber, and a financial crisis in southeast Asia put the brakes on any sales expansion in that region.

With overhead costs soaring and the timber supply dwindling, bold action was required. In mid-July of 1997, the company announced a $47.5-million restructuring and modernization plan: the lumber operations would be consolidated at an upgraded Chasm facility, resulting in the closure of the Exeter and Lillooet sawmills.

Ainsworth's objective was to improve log recovery and reduce its reliance on U.S. markets by producing more value-added products for export. The expansion at Chasm would also give the company more flexibility to produce lumber products that weren't subject to the U.S. quota.

The 28-year-old Exeter mill was pumping out 116 million board feet of premium green-end studs a year, but the operation was no longer competitive. By this stage, older mills like Exeter were becoming a rarity around the province. The fact that Exeter was able to produce as well as it did was a testament to the skill and dedication of the maintenance staff. Many of those charged with keeping Exeter running had been there over 20 years and knew the place inside out. Bob Glover had been site manager for nearly two decades, and the sense of teamwork he instilled at Exeter from the start was a key factor in the mill's longevity.

The "B" shift at the Exeter sawmill in 1997.

Above: Tom and David Ainsworth share a laugh at a company dinner in the mid-1980s. Right: David and Susan Ainsworth were thanked by the town of 100 Mile House on August 5, 2000, the town's 35th anniversary, for their many decades of service to the community and beyond. The ceremonies included the planting of a mountain ash in the couple's honour.

Of the $47.5 million for restructuring, $32 million was earmarked for capital projects at Chasm, which had undergone an $8-million upgrade in 1995. The earlier project included a new computerized log scanning and breakdown system designed to reduce sawdust and improve sawing accuracy. The results showed Chasm was moving in the right direction: the amount of lumber recovered from logs was improved by 20 per cent, and the volume of premium-grade products destined for Japan was increased by 70 per cent.

This newly announced upgrade at Chasm included a 16,000-square-foot plant addition, along with a canter twin-curve sawing line to replace the Chip-N-Saw. Several of the key improvements were designed to increase log recovery by up to 12 per cent. Other changes were included to widen Chasm's range of products, such as lumber produced in metric dimensions for Japan and Pacific Rim markets.

The mill also benefited from a new hot-oil energy system, utilizing wood waste to heat the oil. Plans also called for installing a new kiln and converting the existing trio of drying kilns from hot water to hot oil.

The expansion at Chasm would create 60 new jobs for a total of 265, which would help to offset the 185 positions lost with the Exeter and Lillooet closures.

Construction on the project began in September 1997 and the first boards from the revamped facility were being processed by June 1998. The goal was to produce 225 million board feet of lumber annually.

As part of the restructuring, the specialty plywood plant at Savona and the veneer plant at Lillooet also underwent major improvements to reduce operating costs and boost the production of value-added products.

At Savona, work began immediately on a $9-million, 33,000-square-foot expansion at the specialty plywood plant. New veneer-drying and grading equipment, along with a new

The 1997 expansion project at Chasm included a new canter twin-curve sawing line.

The Savona specialty plywood plant underwent a major expansion.

By 1998, Lillooet was one of North America's leading producers of quality veneer.

pre-press on the High Density Overlay press, enabled production of panels up to 10 feet in length, opening doors to new products and expanded markets.

Producing some 150 million square feet annually, Ainsworth's Savona division would soon become North America's leading supplier of overlaid plywood products used for concrete-forming systems on large construction and urban infrastructure projects. The overlaid forming panels are also used for exterior paint-grade highway signs, and the company's Transdeck panels are used in flat-deck trailers. Savona's production capacity had doubled since Ainsworth took ownership in 1987.

At the same time, Savona's sister plant in Lillooet underwent a $6.5-million improvement project that included a 6,000-square-foot expansion and numerous equipment upgrades. When full production resumed in the spring of 1998, Lillooet had a much faster lathe with increased cutting precision, thereby improving recovery. The project transformed Lillooet into one of the most efficient manufacturers of high-quality veneer in North America.

B.C.'s dismal economic climate in the mid-1990s couldn't have contrasted more sharply with the buoyant state of affairs in neighbouring Alberta. Premier Ralph Klein's pro-business Alberta Advantage campaign was in full swing. It set the tone for how business was conducted in the province, and the Ainsworths welcomed the change. Unlike government officials in B.C., Alberta officials were more interested in how they could be of service.

In matters of resource use, B.C.'s prescriptive forestry code contrasted sharply with a more results-based approach in Alberta.

With the back-to-back successes experienced in 100 Mile and Grande Prairie with new OSB plants, the Ainsworths were looking for ways to sustain the momentum, and Alberta didn't disappoint. In 1996 the provincial government announced a request for proposals on 730,000 cubic metres of aspen and poplar in the High Level area, some 300 miles north of Grande Prairie.

The circumstances around the High Level timber bid differed from the Grande Prairie proposal in several respects. First of all, Ainsworth was no longer a stranger on the Alberta scene. The company now had knowledgeable woodlands staff in the province and was well known among key government officials.

There was minimal competition for the timber this time around, and far more time to prepare the bid. It did, however, involve negotiating timber supply agreements with several small sawmill operations in the area, as well as three sizeable First Nations communities.

Once again, Ainsworth's proposal was headed up by Tim Ryan, with assistance from Dave Cook in Grande Prairie and consultant Carl Leary. Leary, a former boss of Dave Cook's, had 30 years of experience on the northern Alberta scene, making him a valuable asset in the cause.

"At this point we were proposing a much smaller OSB plant at a cost of about $150 million," says Tim Ryan. "That would have made the plant just a bit bigger than Grande Prairie. Then the government made more timber available. By now we had a strong relationship with the High Level community, so with them working with us, we went after that additional timber. We were able to supplement our original bid by another 400,000 cubic metres."

With a projected annual cut of over 1.3 million cubic metres, it was now possible to build what was at the time the world's largest single-line OSB plant. The facility would have a yearly output of over 800,000 square feet of panels — up to a billion square feet at peak capacity.

Unlike the bid in Grande Prairie, the Alberta government was far more lenient about a startup time. They recognized that High Level was an isolated community and that the OSB markets had fallen off in the interim. Rather than stipulating that ground breaking had to commence within 30 days, they were willing to grant several years before any building activity had to take place.

> We can go to six meetings and move things along in negotiations. But if we come to an impasse or we want to make a very key point and set the tone, we say to Allen, "Come on, we need to have you with us." It makes the other parties fully aware that what we say we're doing is exactly in step with what the ownership says, because one of the owners is right there. When people see an Ainsworth, whether it's a contractor or a community person, they say, "Yeah, there's the president of the company and they're showing an interest." In any meeting we go to, it eventually becomes useful to have someone with the last name of Ainsworth with you. The Ainsworths are willing to put their name and face forward for their business. It's a positive; that name helps make the extra bit of difference.
>
> — TIM RYAN

The High Level plant is over 800,000 square feet in size. *One pick with the giant portal crane can empty a bunk of logs.*

High Level was perfect for Ainsworth's expansion strategy in Alberta, although the project did necessitate taking on an equity partner to share in the financing. Ontario-based Grant Forest Products was also interested in expanding its operations and agreed to form a joint venture at High Level. The new company was called Footner Forest Products, with the production runs to be split between both partners' brands.

Grant had been in the OSB business since the early 1980s and operated plants at Englehart and Timmins, Ontario.

The building site chosen for Footner was six miles south of High Level, a busy community of 4,000 inhabitants catering to forestry and the oil and gas industry.

In addition to being the world's largest OSB plant with a production capacity of 860 million square feet annually, Footner's manufacturing process would feature a 12-foot-wide, 187-foot-long continuous press, the largest of its kind in the world used for OSB.

Land clearing for the plant and its mile-long log yard began during the summer of 1999, and the first board rolled out of the press on October 20, 2000. During 1.2-million man-hours of construction activity, not a single lost-time accident was recorded among the several hundred contractors who had descended on the building site.

Footner's newly assembled woodlands staff had the responsibility of bringing in some 800,000 cubic metres of aspen and poplar from a standing start. The task of logging and hauling in northern Alberta is made even more challenging by the short harvesting season. Much of High Level's timber supply area of 4.2 million acres is covered in wetlands; the only time for access is during a window of 100 to 125 days when the ground is frozen solid.

Supplying timber for the mill's startup involved new staff, new contractors, new equipment and a new way of logging to 16.5-foot lengths. Despite some long hours and a steep learning curve, Footner's woodlands staffers had a ready supply of logs in the yard when it came time to start up.

Those initial agreements negotiated with the area's First Nations now form an important part of the plant's timber supply. Roughly one-third of Footner's wood comes from volume supply agreements with the Dene Tha', Askee and Netaskinan First Nations, as well as the Paddle Prairie Metis Settlement.

Within several years, the company had also initiated Alberta's first joint Forest Management Agreement (FMA). It took two years of negotiation, but Footner and Tolko now share the same resource base and manage it cooperatively — Footner for the deciduous stands and Tolko for the softwood stands for its sawmill. The agreement enhances Footner's tenure rights to the timber, as well as providing a framework for sustainable management.

By the time Footner Forest Products began producing board, Ainsworth had become a significant player in Canada's OSB industry. In 2001 the company had sales of over $310 million, producing over 1.4 billion square feet of OSB between their three operations.

The growth, however, came at a price.

The commitments to High Level had left the company in a vulnerable financial position. Stronger markets would have lessened the impact, but an over-supply of oriented strand board was pushing prices down. In 2001 OSB prices were down 22 per cent from the previous year, and export sales to Japan had gone flat.

This wasn't the first time Ainsworth Lumber had found itself in a struggle, but the stakes were far higher than they'd ever been in nearly five decades of business.

With the company looking to reduce its overhead, tough decisions had to be made on a number of fronts. Capital spending was trimmed back and several major expansion projects were abandoned or put on hold, including a second production line at Grande Prairie. The site preparation there had already begun in the summer of 2000. Plans called for a new line to manufacture a value-added "combi-panel" consisting of an OSB core with particle-board surfaces on the top and bottom.

High Level's 12-foot-wide, 187-foot-long continuous press.

A continuous mat of OSB is ready to move through the press.

Back in B.C., the company decided to sell the Chasm division sawmill. Forestry giant West Fraser eventually purchased the assets and two timber licences in a deal that was announced in the spring of 2001.

This left the Abbotsford finger-joining plant as Ainsworth's sole producer of lumber. Without Chasm's steady shipments of trim blocks as raw material, Abbotsford was forced to look for other suppliers at considerably higher prices. The softwood lumber dispute continued to drag down the markets, and currency fluctuations cut margins even further.

Given the conditions, production of finger-joined lumber had to be scaled back in the fall of 2001, with a full curtailment following a few months later. The Abbotsford plant was officially closed November 28, 2004.

Despite the uncertain times, the company's OSB facilities continued to improve their performance. Almost every quarter, Ainsworth employees at mills in 100 Mile and Grande Prairie set new production and efficiency records. For the first time in company history, the combined output of both mills exceeded 1.25 billion square feet of OSB. To reach this milestone, both plants had been surpassing their design capacities for several years, making them among the most efficient operations in the entire industry.

By 2002, it was clear that the company's situation had begun to turn around. Prices for OSB had actually decreased a further two per cent, but Ainsworth had increased their sales by nearly $30 million.

Heading into 2003, the company's prospects appeared somewhat more positive than in the previous year, but no one could have predicted just how much their fortunes would shift. Fuelled by low interest rates and pent-up demand, the U.S. housing construction market was at its busiest level in 25 years. As a result, OSB prices were up nearly 46 per cent over 2002. By October the markets were at an all-time high.

The turnaround for Ainsworth couldn't have been more dramatic. The company's OSB sales were up over 50 per cent, totalling nearly $400 million. Between 100 Mile, Grande

A NEW BRAND IS UNVEILED: *Ainsworth*Engineered®

In 2000 the company launched its new *Ainsworth*Engineered brand as a tangible symbol of its focus on higher-quality products rather than basic commodities.

"The new brand gave us an opportunity to draw out what we feel is our company strength: the Ainsworth culture of quality," says Robert Fouquet, the company's vice-president of marketing.

"There's an incredible level of commitment that Ainsworth employees show in producing and delivering the highest-quality products possible. We wanted something to accurately reflect that."

The brand is well-suited to the company's emphasis on developing new value-added components for engineered wood housing packages. It also implies that Ainsworth is "engineering its future" rather than being at the whim of market forces.

David Ainsworth commented at the time, "I'm totally enthused about the *Ainsworth*Engineered brand. I think it's a wonderful idea. Sometimes these younger guys come up with some pretty good ideas — you kind of wish you had thought of it yourself!"

Prairie and High Level, Ainsworth produced 1.35 billion square feet of OSB — up more than five per cent from the previous year.

And the gains weren't limited to OSB. Savona's specialty plywood division, also riding the wave of increased construction activity in the U.S., saw its sales increase by over 21 per cent. The OSB and plywood divisions combined for a record total of $480 million in sales — an increase of 35 per cent over the previous year.

Leaving lumber production behind and pegging the company's fortunes to OSB had turned out to be a wise course of action. The unprecedented markets of 2003 and 2004 couldn't have come at a better time. Buoyed by record sales, the Ainsworths were once again able to look at expanding their operations.

Before the end of 2003, Michael Ainsworth heard that the Voyageur Panel OSB mill at Barwick, Ontario, might be put up for sale.

Although the plant was jointly owned by Boise Cascade, Abitibi Consolidated, Northwest Mutual and Allstate Insurance Company, it was operated by Boise Cascade.

"I knew a person at Boise Cascade through the APA (Engineered Wood Association), so I called him up and said, 'It may not seem real obvious to you right at the moment, but I want to make sure that if you do decide to sell that mill, please let us know.' He said they hadn't decided yet if they were going to, but if and when they did, it would be an open and public process, and sure, they would put us on the list," says Michael.

"Sure enough, about that January or so they did decide to sell. As it turned out, we were in the final stages of refinancing the company, which meant the removal of any restrictions that would have prevented us from making a bid on the mill."

By mid-April 2004, Ainsworth was signing a deal to purchase the Barwick plant for $216 million (U.S.). The acquisition increased Ainsworth's total OSB production capacity to over 2 billion square feet annually and, most importantly, extended its reach into the eastern markets. An additional 150 employees brought the company's total payroll to nearly 1,200.

The Barwick plant — slightly larger than 100 Mile's — was built in 1997. It was producing over 440,000 square feet of OSB annually when Ainsworth took ownership. Less than a year later, that figure was up to 480 million square feet with almost no capital expenditures.

The OSB markets were showing considerable staying power into 2004, and Ainsworth was in the fortunate position of having its production capacity increase by nearly 25 per cent. With a modern, well-run plant in Ontario now part of the fold, Ainsworth had the critical mass to look at opportunities that would have seemed unimaginable just 18 months before.

The ink was barely dry on the Barwick acquisition when the Ainsworths started going over the numbers on an even bigger deal, one that would solidify their presence in eastern markets and take them international at the same time.

> *Except for learning patience, where to look for information, or how to pay attention, none of the stuff you learn at university helps you do this job. You're sitting in a business meeting sucking it all up like a sponge. They say something and you say, "What do you mean by that?" and you get them to explain it to you. Every time you go to these meetings there's another acronym, and you go, "What the heck does that mean?" Or someone will mention that maybe you could look at this market or that market, and you say, "What is that market?" and you learn more about it. Then, after having done two or three or four financings, after a while you learn the different terms and the ins and outs, and then you have to come back and make them relevant to your own business.*
>
> *But you just ask a billion questions. How do you learn about all the different equipment in an OSB plant? Meeting after meeting after meeting. You make notes, you ask every stupid question in the book. The finance-accounting side is no different. You ask stupid questions all day long until you get enough information to understand.*
>
> — MICHAEL AINSWORTH

Acting through an intermediary from New York–based Deutsche Bank Securities, the company had put out feelers to Potlatch Corporation to find out whether they'd consider selling their three OSB plants in northern Minnesota.

Potlatch took Ainsworth up on its offer, and by August 2004 the two companies had signed a definitive agreement. By mid-September Ainsworth had closed a deal to purchase the Potlatch assets for $455 million (U.S.), a transaction that vaulted them into the position of fourth-largest producer of OSB in the world.

The company now owns OSB divisions in Bemidji, Grand Rapids and Cook, Minnesota. The U.S. plants add an additional 1.3 billion square feet of OSB to Ainsworth's production capacity, and 600 new employees — all under the banner of Ainsworth USA LLC, the company's legal name in the United States.

The Minnesota operations differ from the company's Canadian holdings in two important respects. First, all three U.S. mills are generally older. The mills at Bemidji and Cook were built in the early 1980s, with Bemidji being one of the first OSB plants in North America. A second production line was installed in 1990, making it the largest of the three divisions. The Cook facility, meanwhile, underwent a complete upgrade in 2000.

Dad always had the ability to see an opportunity or simply create one if necessary. And he did this throughout the company's history — often in spite of overwhelming obstacles at any given time.

In the early 1960s there was pressure from others who didn't want to see anyone else succeed as a competitor in a new industry. When Dad was unable to secure a sufficient and continuing supply of Douglas fir logs, he set out to develop a business producing lumber from lodgepole pine.

Other producers had a sufficient supply of Douglas fir logs and had no requirement to harvest lodgepole pine. At the same time, however, they took steps to deny Ainsworth Lumber's access to the pine — even though they didn't need it or couldn't even use it themselves.

Producing lumber from pine was virtually unheard of, so Dad went ahead and pioneered small-log sawing as the means of production. Now he was producing kiln-dried precision end-trim studs, but there was no established market for his new product. So, in fact, he developed the market as well.

As part of a three-member sales cooperative, he was among the first producers in the Interior to treat his lumber with a distinctive green end-paint. Before long, those studs would become a much larger segment of the construction market than the original Douglas fir lumber.

In the 1980s, by the time the pine industry was fully developed and mature, he was already looking for the next step. That's what led him to using aspen and residual pine for oriented strand board. He felt strongly that OSB was the way to go, despite well-meaning advice by others to never go near that segment of the industry.

In so many instances, he just went out and created these opportunities.

Without a doubt, the real value in our company now is engineered wood products or OSB. This never would have happened at Ainsworth without Dad's doggedness and determination. And while all the rest of us did the heavy lifting, managing and running the company around that time, it was Dad's vision that got us into OSB. It wouldn't have happened without him.

Thanks to that vision, brought to fruition in so many ways over the years, the company has an incredible future.

— ALLEN AINSWORTH

Ainsworth acquired the OSB plant in Barwick, Ontario, in the spring of 2004.

The Grand Rapids operation was actually built in the early 1970s, when it produced waferboard. It underwent a retooling in 1985 to produce OSB.

Within their first year of ownership, Ainsworth committed substantial resources for long-overdue capital improvement projects, particularly at the Bemidji and Grand Rapids facilities.

Second, the U.S. mills rely on a different timber procurement procedure. While the Canadian system utilizes long-term licences and management agreements on Crown land, as well as private sales, the Minnesota supply is far more diverse. In the U.S., mills acquire their timber through sales and auctions where prices are fluctuating constantly. Ainsworth's U.S. timber buyers work with some 300 suppliers at any given time in a market-driven system, with logs coming in from federal, state, county and privately owned land. The three mills process a total of 1.85 million cubic metres of wood each year.

For a relatively small company, in OSB for barely 10 years, the events of 2004 were phenomenal. With the Ontario and Minnesota acquisitions, Ainsworth doubled in size in a period of six months. The company's total annual production capacity now exceeds 3.3 billion square feet. Annual sales are in excess of $1 billion — with over 90 per cent of revenues from OSB — and there are approximately 1,800 employees on the payroll.

For some departments, the implications of doubling in size were immediate. Ainsworth's traffic department in Vancouver was suddenly responsible for overseeing the timely delivery of twice as much product as before. It took a few all-nighters for staff to sort out the logistics, but not a single shipment was missed during a very hectic transition period, and the department now oversees the movement of some 20,000 railcars and 40,000 transport trucks in the course of a year.

This doubling of production capacity has spurred other developments as well.

By the spring of 2005, Ainsworth was finally in a position to dust off its plans for a major expansion at the Grande Prairie division. The project which had been put on hold in 2000, and kept alive by several hard-won extensions to the original fibre-supply agreement negotiated in the late 1990s, is once again in the forefront.

When the $300-million expansion is completed, the existing plant at Grande Prairie will double in size and have the ability to produce an additional 600 million square feet of OSB annually. The entire mill will cover 22 acres, making it the largest OSB plant in the world. Taking advantage of the latest technology, the facility will feature an 8½-foot-wide, 180-foot-long continuous press that facilitates production of extra-thick panels known as laminated strand lumber, or LSL, as well as OSB.

The expansion will also create 85 new jobs at the plant, in addition to 200 jobs in the surrounding area.

The company has also committed to building a remanufacturing plant for rimboard and oriented strand lumber in Valleyview, about 60 miles east of Grande Prairie.

When Ainsworth started producing OSB at its first mill in 100 Mile, the whole industry was focused on commodity production and market share in competition with plywood. Nobody in the world of panel sales believed customers would pay a premium price for a higher-quality OSB panel.

Another load of logs arrives at the Grand Rapids facility in Minnesota.

OSB sales manager Blair Magnuson recalls those early days of production and how Ainsworth entered the marketplace.

"I wrote Allen a big memo about the quality attitude and the fact everybody told you that you couldn't get a premium for a quality product," says Magnuson. "We just forged ahead and mainly had the goal of quality right from the start. We found out you could do some things to improve the quality of the product and that people would pay a little more. We were a company that didn't have any salespeople who were OSB panel people, we just focused on quality and never put a stick on the ground from day one; we didn't have to be putting inventory outside. We could sell our product because it was good quality and we had good relationships with our customers. It wasn't long before we were asking for letters from our customers, asking them about what they thought of the product. We got great responses."

Within a year of startup, 100 Mile's OSB panels were certified for the Japanese market — proof that they had met the toughest quality standards in the industry.

Starting out with sales of 40 million square feet to Japan, Ainsworth captured 60 per cent of the Japanese market within five years, and enjoyed sales totalling 160 million square feet. By 1997, Grande Prairie, with its 12-foot-wide production line, was also certified to produce the specially sized JAS-quality panels.

The painstaking process years earlier of meeting the Japanese standards for quality in lumber products had been well worth the effort: virtually every Japanese customer buying Ainsworth OSB had started out buying the company's turquoise-end studs. By the time Ainsworth OSB was on the market, featuring a distinctive turquoise edge-seal, the Japanese were well aware of the company's reputation for quality products.

The Bemidji plant is the largest of the three Minnesota OSB operations acquired by Ainsworth.

The OSB facility in Cook, Minnesota, underwent a substantial upgrade in 2000.

The company's offshore sales are managed through a consortium called Interex Forest Products, based in Vancouver. Ainsworth is one of six western Canadian forest companies that form Interex.

Ainsworth's drive to develop new value-added products and then generate the markets for them requires a good deal of time, talent and resources. From its initial entry into OSB, Ainsworth has put particular emphasis on building its bench strength throughout the company by attracting top-notch people for research and development, process optimization and a host of other positions that all contribute to keeping Ainsworth on the leading edge.

New products in particular can involve months and sometimes years of testing, either in-house or at independent facilities. Ainsworth has made extensive use of partnerships and joint projects with university research departments, as well as with a wide range of government research centres and programs.

Ainsworth's line of value-added products has gained wide recognition in the marketplace. Over the past several years, the company has introduced SteadiTred, used in residential stair systems; Durastrand flooring, a premium-quality tongue-and-groove flooring product; Durastrand rimboard, used in stair stringers and short-span headers; and Thermastrand Radiant Barrier panels, which reflect the sun's rays in hot climates.

The company's various webstock products are used to produce engineered I-joists. Ainsworth webstock is sold to at least 10 different manufacturers of I-joists, all of which require separate specifications for thickness, strength and panel size.

> *It's all about making sure we're running our business the best that we know how, and it all comes from David. We've got 1,800 people and we're trying to manage this company so that we're still able to sit down with a maintenance manager, or any employee, and talk about what it is they're doing and why they're doing it. Even though we're a bigger company, we still believe we need to have a better product, and we've got to service the heck out of the customers. And there's no question that comes from David, absolutely.*
>
> — MICHAEL AINSWORTH

Foil barrier is used in Ainsworth's Thermastrand panels. *Another bundle awaits delivery at Barwick.*

Whether it's value-added products or commodity sheathing, current trends in OSB sales indicate a bright future. Analysts predict that OSB's market share over plywood will rise to 80 per cent by 2013. That's up from a current level of 60 per cent market share for OSB panels.

As the OSB industry matures, a new product called laminated strand lumber, or LSL, is destined to play a significant role, and Ainsworth is poised to be a front-runner in its production. Manufacturing the panels at thicknesses up to 1¾ inches requires state-of-the-art technology and the production expertise to go with it. The new forming line and continuous press at Grande Prairie's second line is specifically designed with LSL manufacturing in mind, although production will begin with traditional OSB products.

In the meantime, Ainsworth's strong reputation in the marketplace can be traced right back to their first mill and the way it was operated and maintained.

First and foremost, the company is among the safest in the industry. Ainsworth made a conscious decision to revitalize its employee safety program in the early 1990s. By 2002 its accident frequency ratings were well below the forest industry average. More recently, Ainsworth has surveyed its employees company-wide to fine-tune its safety and wellness program even further, garnering a number of top industry and government awards along the way.

Several divisions, as well as woodlands, have now gone for up to two years without a single lost-time accident.

When it comes to the production capabilities of Ainsworth's mills, the quest for improvement never ends. 100 Mile OSB became certified under the world-recognized

ISO-9001 standard in 2000, which involves monitoring every aspect of the production process. More recently, the woodlands operations at Grande Prairie were certified under the ISO-14001 standard, which focuses on environmental practices. B.C. operations were certified under the same program in 2004.

Public concern about the environment has also made it important for companies like Ainsworth to demonstrate sustainable forest management practices. In addition to the ISO-14001 certifications, Ainsworth's B.C. and Alberta woodlands divisions have also earned their certification under the Sustainable Forest Management (SFM) program, and Grande Prairie is certified under the province's Forest Care standard. In the U.S., Ainsworth's Minnesota woodlands operations have recently been certified under both the Sustainable Forestry Initiative (SFI) and Forest Stewardship Council (FSC) programs.

Qualifying for most of these operational and environmental standards involves months and often years of preparation, followed by intensive independent audits and follow-up audits each year thereafter.

The Savona division, meanwhile, recently achieved the SFM "chain of custody" certification, which guarantees that each panel produced comes from logs that were harvested according to sustainable forestry standards set out by the Canadian Standards Association.

The company also contributes substantial resources to a wide range of studies and long-term research projects covering everything from grizzly bear habitat to ecosystem mapping in Canada's boreal forests.

Silviculture is another important aspect of Ainsworth's ongoing forest management program. The company has been involved in several research projects designed to identify the best hybrid poplar species for tree plantations in Alberta. Near the Grande Prairie facility, work has already begun on Ainsworth's first full-scale tree plantation. In the right conditions, deciduous tree farms could provide an important future supply of fibre for the company's OSB mills.

> *Our mantra has always been "all or nothing." Every family member agrees or we don't do it. There have been some pretty interesting discussions around that, but at the end of the day everybody agrees on a direction before we go ahead.*
>
> *Getting bigger is going to present a whole new set of challenges, but we'll get through them just fine. It's all about having good people and having them share a sense of ownership in the business. I don't think it will be different than any other growth we've had. We didn't even know how to make OSB 12 years ago. I'm pretty confident we're in the top percentile of companies out there now and that we're making quality OSB. Our products have an excellent reputation, our mills run well, and we have good people.*
>
> — DOUGLAS AINSWORTH

For the communities that Ainsworth calls home, the company has been a steady supporter of countless programs and activities through donations large and small. In Grande Prairie, the David and Susan Ainsworth Fund has contributed more than $1 million since 1995.

More recently, the company took the unprecedented step of sponsoring the construction of low-cost housing in hurricane-ravaged Louisiana. Working with the world-renowned Habitat for Humanity program, Ainsworth contributed over $150,000 (U.S.) in 2006, which enabled a number of families to gain a fresh start in life.

Over 50 years have passed since David Ainsworth first rolled through 100 Mile with his portable sawmill in tow, and the company he and Susan founded is still a work in progress. Plans for expansion in B.C., Manitoba and the eastern U.S. are on the drawing board, and if the past is any indication, further opportunities are sure to arise.

Woodlands general manager Dave Cook (holding wagon) represented the company in a special ceremony to acknowledge Ainsworth's contributions to over 200 worthy causes in the Grande Prairie area since 1995. The donations were given on behalf of the David and Susan Ainsworth Fund. The red wagon, awarded by the Alberta Promise program, symbolizes "pulling together for children and youth." Presenting Ainsworth with the award were (from left) Heather Forsyth, Minister of Children's Services, Premier Ralph Klein and his wife Colleen, honorary chair of Alberta Promise.

"We're builders, we're not coasters," says Brian Ainsworth. "Our life has just gone along putting one foot in front of the other, right from the start. There are no plateaus that we get to and say, 'We've made it.' If there was just one generation involved, there would be plateaus. But you always need another generation to keep moving forward."

Staying in close touch with a growing company spread across several provinces and two countries has its challenges, but the Ainsworths are determined to remain hands-on owners. The members of the current generation — Catherine, Michael, Douglas and Kevin — are well aware that the company's future hinges increasingly on the decisions they make. And in an industry where markets and conditions can turn on a dime, they don't take success for granted.

Over the years they've immersed themselves in the details of design, construction and efficient operation of their facilities. The family members each have their own roles and responsibilities, but there's a considerable amount of shared expertise at their disposal when major decisions are on the table, whether these involve plant upgrades or the intricacies of corporate finance.

Succession plans in any company, whether publicly or privately held, are a crucial part of long-term viability, and Ainsworth has taken steps to ensure a smooth transition.

"The third generation has spent literally 20 years in the trenches doing a variety of jobs throughout the company," says Allen Ainsworth. "Now they've been given very specific roles, with Brian and me and our parents mentoring them along. In the past couple of years they have very purposefully traded off responsibilities in order to more completely learn the skills in the other disciplines of the business."

Although still in the formative stages of their careers, three of Catherine and Hanns's four children also work at Ainsworth: Susan, Ryan and Steven Fellenz.

The combination of successive generations of owners, employees who share the vision, and a deep determination to keep moving forward has made Ainsworth Lumber the dynamic enterprise it is today. David Ainsworth's steady resolve has showed time and again that no problem is insurmountable, and his passion for quality and innovation has set the tone for the company since the 1950s. Over 50 years later, those same attributes continue to define Ainsworth Lumber and its reputation in the industry.

Etched in glass in the company's corporate boardroom are larger-than-life images of David and Susan Ainsworth, depicting some of their earliest days in the Cariboo. It's a fitting tribute to these two individuals and provides an enduring connection to the company's past.

Their presence is sure to be felt for many years to come.

Three generations of the Ainsworth family. Back row, from left: Susan Fellenz, Douglas, Kevin, Michael and Catherine Ainsworth. Front row, from left: Allen, David, Susan and Brian Ainsworth.

LOGGING TERMS

Barber chair: A tree that has split during falling, usually as a result of improper cutting.

Bean burner: A camp cook, usually not a good one.

Blowdown: Trees that have been blown down by wind.

Bull buck or bull bucker: A supervisor of fallers.

Bullcook: A man who cuts wood, cleans up bunkhouses and does assorted duties around the logging camp.

Bull of the woods: A logging superintendent.

Bush ape: Someone who works in the woods.

Cant: A partly trimmed log.

Cat choker: A choking cable for logs with an eye at one end.

Cat face: A scar or deformed section at the base of a tree caused by fire or rot.

Cat skinner: A bulldozer operator.

Chaser: A worker who unhooks chokers at the landing.

Choker: A short length of wire rope, one end of which is noosed around the log while the other is attached to the main line or butt rigging.

Cruiser: A worker who inventories volume and grade of standing timber.

Crummy: A car or bus used to transport logging crews to work area. In railroad shows, an enclosed conveyance used for the same purpose.

Domino falling: Placing under-cuts and back-cuts in a series of trees, then pushing them over with another tree. Considered dangerous and against safety regulations.

Don't break suction on the seat or you're in trouble: Warning to an equipment operator about to descend a steep or slippery slope.

Gut bucket: A logger's lunch pail.

Gyppo: A fly-by-night operator, or one who operates on a shoestring budget.

Hangup: When a tree becomes lodged in another and can't fall to the ground.

Haywire show: A logging operation without safety standards.

Highball: To go ahead fast.

Hobo log: An unattached log that is picked up by a passing skidder already loaded.

Iron: A catchall phrase for modern logging equipment, from skidders to feller bunchers.

Iron on rubber: When heavy logging equipment is transported on low-bed trucks.

Jackstrawed: Trees or logs piled in a disorderly manner.

Jersey cream: Especially good timber.

Keep the black side down: Keep the logging truck's tires on the road.

Log wrench: A peavey or hooked pole used to move logs by hand.

Long sticks: Long logs that exceed the boundaries of a landing.

Nosebag show: A camp where a lunch bucket is carried.

Nut splitter: A locomotive mechanic.

Peeler: A log of sufficient diameter to peel and use as veneer for production of plywood.

Powder monkey: A worker who uses dynamite in a logging operation.

Put the jewelry on: Put chains on the truck's tires.

Scaler: A worker who measures each bucked log for number of board feet and grade.

School marm: A tree stem that branches into two or more trunks or tops.

Show: The area being logged.

Side: A crew and equipment. Can mean complete yarding, loading, falling and bucking crew.

Sidewinder: A limb or sapling that is dangerous to cut because it is bent under a felled tree.

Skyline: A heavy cable hung between two spar trees with a travelling carriage to haul logs through the air in rough country.

Slab orchard: A sawmill.

Slash: A logged-off area. Also means to cut a line through bush for a survey crew.

Snag: A dead tree.

Springboard: A board old-time fallers stood on to fall trees.

Stagged trousers: Pants cut short to prevent tripping hazards.

Suicide show: A dangerous piece of ground to log.

Tame ape: Someone who works in a sawmill.

Tin pants: Water-repellant trousers.

Widow maker: Loose overhead debris such as limbs or tree tops that could fall at any time.

ACKNOWLEDGEMENTS

Producing a book like this would be an impossible task without the participation and assistance of many individuals.

I'd like to acknowledge the nearly 50 people who were kind enough to be interviewed — many of them more than once — including Ainsworth family members and relatives, employees past and present, contractors and various friends of the company. Efforts on the part of many individuals in locating photographs were also greatly appreciated.

Valuable background information on B.C.'s forest industry came from a variety of sources, including several books by Ken Drushka: *Tie Hackers to Timber Harvesters; Working in the Woods; and Lignum, A History*.

Two Ainsworth anniversary supplements produced in 1977 and 1992 by the *100 Mile House Free Press* were useful sources of information as well. Doris Cordel at the Halkirk Visitors Centre was very helpful with information about the town's formative years.

My thanks also go to Al Smith and Julia Derek for their assistance along the way.

Paul Luft
100 Mile House, 2007

INDEX

Page numbers of photos are given in italics. Members of the Ainsworth family are identified in relation to David Ainsworth (DA) and Susan Ainsworth (SA).

Abbotsford division 214–17, 221
   closure 275
   production 265
   Vancouver sales office 240, *243*
Adams, Bruce *194*
aerial surveys 159, *161,* 182
Ainsworth, Allen (son of DA and SA) 23, *49, 79, 101, 206, 211, 230, 237, 285*
   Chasm mill 148–49
   childhood *40,* 41, 58, *58,* 61, 62, 70, 72, 73–74, 83, 87
   children 118
   CP Air Flight 21 173
   DA as visionary 277
   Evans Forest Products 219
   financing OSB 240–43
   finger-joined lumber 189
   flying 161, 163, 171, 214
   Grande Prairie OSB 255, 256, 258, *262*
   joins AL 70, 78, *81,* 105–7
   Little Bridge Creek Logging 107
   management 206, 207
   marriage 107
   Mountain Pine Lumber 133
   100 Mile OSB plant opening *252*
   sales and marketing 157
   stud mill upgrade 129, 130
   succession 285
Ainsworth, Brian (son of DA and SA) 41, *49, 79,* 105, *137, 206, 230, 237, 285*
   birth of son 118
   childhood 56, *58,* 62, 70, 72, 73, 83, 87
   Class 1 ticket 114
   company philosophy 284
   CP Air Flight 21 173
   flying 161
   green airplane hangar 103
   joins AL 70, 78, 105–6, 107
   Little Bridge Creek Logging 107
   management 206, 207
   marriage 118
   100 Mile OSB plant opening *253*
   roadside logging 139–40
   stud mill upgrade 129
   woodlands 157
Ainsworth, Catherine (daughter of DA and SA) 61, 87, 107, *230, 237, 241, 285*

   children 285
   joins AL 157, 209–10, *209*
   marriage 210
   100 Mile OSB plant opening *253*
Ainsworth, David (DA) *38, 65, 67, 74, 105, 194, 206, 230, 237, 270, 285*
   AinsworthEngineered 275
   birth 16, 17
   Cariboo 52–69, *53,* 72–79, 81
   childhood *17, 18,* 17–20, 22
   children 23
   Cowichan Valley 23, *26, 29,* 31, 33–36, *35,* 39–41, *40*
   CP Air Flight 21 173
   employees 64, 78, 83, 117, 118, 181, 183, 197, 250
   flying 159, *164, 179,* 180–82
      cattle rustler 171
      emergency flights 163–65, 167, 168–69, 183
      first plane 161
      flying with grandchildren 178–79
      helicopters 162–63, *162, 163,* 168, 170, *174,* 175–76, *177*
      jet aircraft rating 165, *166,* 167, *167, 180*
      pilot's licence 160–61
      search and rescue 174, 176, 182, 184, *184–85*
   gasification plant 150–52
   hands-on approach 83
   independent contractor 41–42, 45
   innovation 66–69, 115, 132, 224–25
   lodgepole pine 91–97, 100–4, 111–14
   logging injuries 46
   marriage 22, *23*
   mechanical abilities 19, 55, 64, 73–74, 85, 121
   move to 100 Mile House 86
   100 Mile House community 86, 89, 157, *238, 270*
   100 Mile OSB plant 244, *244,* 251–52, *252, 253*
   oriented strand board 225, 227, 228–32
   PA-16 hearings 235–37, *235*
   portable sawmill 11, 47–48, *48,* 50–51
   respect for equipment 64, 66, 73, 132, 143
   Tom's departure 57–58
   visionary 118, 277
   working round the clock 66, 69, 85, 104, 180
Ainsworth, Diane (niece of DA) 53
Ainsworth, Douglas (grandson of DA and SA) 118, *163,* 285
   childhood 210, *211,* 212
   DA's portable sawmill 217
   family business 212, 283
   Grande Prairie OSB 256, 260
   joins Exco Industries *213,* 214
Ainsworth, Eiko Uyeyama (daughter-in-law of DA and SA) 118, 157

Ainsworth, Ellen Shelbourne (mother of DA) *12, 13, 14, 15,* 16
    children 15–16, 17
    moves to Canada 12–14
Ainsworth, George Edwin (father of DA) *12, 14, 15,* 16
    blacksmithing *13,* 14–15, 17, 19
    children 15–16, 17
    "Englishness" 15–16
    farming 14, *16,* 17, 20
    moves to Canada 12–14
Ainsworth, Hiroko Uyeyama (daughter-in-law of DA and SA) 107, 157
Ainsworth, John Halkirk (brother of DA) 15, *15,* 17, *18*
Ainsworth, Kevin (grandson of DA and SA) 118, *285*
    childhood 210–11, *211,* 212, *213*
    education (forestry) 214
    flying 178–79
    100 Mile OSB plant 244, *244, 245, 253*
    woodlands 268
Ainsworth, Lana (niece of DA) 53, *49*
Ainsworth, Lee (nephew of DA) 34, *49,* 53
Ainsworth, Michael (grandson of DA and SA) 118, *285*
    Catherine Ainsworth and computers 209
    childhood 210, *211,* 212, *213,* 214
    DA's influence 281
    education (marketing) 211–12, 214
    financing OSB 240–43, *241*
    Grande Prairie OSB 256
    involved in Evans purchase 219
    learning on the job 276
    100 Mile OSB plant 244
    100 Mile OSB plant opening *253*
    studies Japanese market 239
Ainsworth, Muriel (sister-in-law of DA) 53
Ainsworth, Susan (SA) 53, *285, 237*
    Cariboo 57, *58,* 60–61, 62, 69–70, 71, 72, 77
    childhood 20–22, *21*
    children 23, 53–54, 57, 70, 73
    Cowichan Valley *40,* 41, 42
    DA's flying 160, 161, 168–69
    Garry Babcock 145, *145*
    grandchildren 211, 212
    helicopter lesson *168*
    marriage 22, *23*
    move to 100 Mile House 86
    moving cabins 71, *71*
    100 Mile House anniversary *270*
    100 Mile OSB plant opening 252, *252, 253*
    role at AL 73, 77, 83, 84, 115, 117–18, 156, 157, 209
Ainsworth, Tom (brother of DA) 16, 17, 19, *38, 49, 270*
    Cowichan Valley 23, 26, *30, 32,* 34, *37,* 41
    Exeter saw-filing shop *154*
    independent contractor 41–42, 45, 47–48, 50–51
    logging injuries 34, 46
    move to Cariboo 52–56
    rejoins DA 133–34
    returns to Fraser Valley 57
    Second World War 41
Ainsworth Lumber (AL)
    airplanes 159, 175, 178, 180, 185
    board of directors 243
    Clinton timber sale 146–47 (*see also* Chasm division)

community relations 157, 163–65, 203–4, 225, 234–35, 236, 237, *238,* 255–56, 257, 283, *284*
competition from other companies 90, 94–97, 100, 112, 188, 233, 236–37, 277
computerization 209–10
contractors 135–36, 138–39
cost of logs 90, 139, 256, 267–68, 269
employee relations 83, 115, 117, 118, 157, 201 (*see also* Cariboo Woodworkers Association)
expansion into U.S. 277–78
family business 188, 272
financing 100 Mile OSB plant 238–43
finger-joined lumber 189–93
First Nations 224, 236, 250, 256, 257, 262, 268, 272, 274
40th anniversary *238*
gold hard hats 109
green ends *97,* 100–1, 103–4, 155
incorporation 84
innovation 82, 136, 191–92, 193
IWA 201, 205
lodgepole pine 90–97, 101–3, 104, 107–14
logging trucks 107, *114, 196,* 199–201
logo 155
management style 136, 191, 205–7, 222
market fluctuations 202–4, 239–40, 269, 274–75
Mt. Timothy Ski Hill 225
108 Mile Heritage Site 11
oriented strand board 230–32
PA-16 hearings 233–37
pouring foundations 102, 111, 126, 149, 244, *246*
publicly traded company 241–43
quality 82, 100–1, 103–4, 116, 117, 123, 194–96, 222, 250, 275, 279–82
research 281–82, 283
road building 143–45
safety 273, 282
sales and marketing 133, 196–97 (*see also* Ainsworth-Weier's Lumber Sales, Fibreco Exports, Interex Forest Products, Mountain Pine Lumber, Seaboard Lumber Sales)
shipping product 83, 122–23, 152, 198, 199–201
succession 118, 284–85
Susan and David's partnership 78, *116,* 118, *123,* 156
sustainable forest practices 257, 283
timber supply 87, 90–91, 96, 188, 229, 234, 235 (*see also* Pulpwood Agreement No. 16)
value-added products 281–82
wood waste, reduced 64, 73, 119, 126, 153, 189, 270
wood waste, using 101, 103, 120–21, 150–51, 189, 217, 225, 230, *231* (*see also* finger-joined lumber, oriented strand board, wood chips)
wood chips 120–23, 156
See also Abbotsford division, Barwick OSB plant, Chasm division, Footner Forest Products, Grande Prairie OSB plant, High Level OSB plant, Lillooet division, Lone Butte mill, Minnesota OSB plants, 100 Mile OSB plant, Savona division
Ainsworth USA LLC 277–78
AinsworthEngineered 275
Ainsworth-Weier's Lumber Sales 196–97
Alain, Paul 223

Alberta
    forest industry 140, *254*, 256, 260–61, *263*, 272–73
    OSB 227–28, 231, 232, 255
    See also forest industry, Grande Prairie OSB plant, High Level OSB plant
Alberta Advantage campaign 272
Alexander, Jack 171
Anderson, Gus 62
Andreas Stihl factory 31
Armeneau, Harold 66
Armstrong, Lee 245, 251
Asian markets 269, 270 (*see also* Japanese market)
aspen *254*
    in Cariboo *228*, 230
    in OSB 231, 232, 250–251
Auld, Herb 60, *60*, 65

Babcock, Garry 129, 137, 139, 145, *145*
Bach, Art 74
Bach Mobile Sawmill 74, *75*
Baker, Johnny 52
Bambi bucket 170, *170*
Barford, John 245, 251
Barkerville 86
Barko 450 butt 'n top loader 141
Barrick, Don 139
Barwick OSB plant 276, *278*, 282
Battle and Houghland Ltd. 91, 102, 133
Bauer, Gerry 258
BC Rail 197
Becker, Bill 129, 157
Bell helicopters 162, *162, 163,* 170, *170, 174, 181*
Bemidji OSB division 277–78, *280*
Bentley, Peter 120
Bickle, Roy 262
Big Bar ferry 168–69
Bingham and Hobbs Equipment 47
birch 251
Black, Jack 198
Blacklock, Kim 214
Blackstock, Al 85, 102
Blackwater Timber 94
Blais, Gerry 139
Blomfeldt, Len 136
Bodman, Phil 174
Boise Cascade 276
Bold, Heinz 167
Bolha, Phillip *206*
Bond Bros. 129
Bowman, Oliver 46
Bowman, Orion 46
Brandram-Henderson 103–4
Brett, Earl 45–46, 47
Bridge Creek 215
Bridge Creek Estate 79, 86
British Columbia forest industry 59, *105*, 201, 203, 239–40, 265–66
    B.C. Interior 61–62, 119–20, 146, 203, 234
    Cariboo 87, 90, 114–15, 119–20
    conservation 187, 223, 239–40, 267
    consolidation 97, 100, 120, 125, 156, 188

Cowichan Valley 25–27, *31, 32,* 34–37, 39
    logging regulations 119–20, 139–140, 143, 145, 146, 187–88, 266–67
    stumpage 90, 110, 223, 240, 267
    timber allocation 62, 90–91, 94–97, 141, 143, 188
    wages 33, 46, 115, 137
    See also British Columbia Ministry of Forests, Commission on Resources and Environment, environmentalism, forest industry, softwood lumber, *specific laws*
British Columbia Ministry of Forests
    AL's OSB proposal 230, 232–33
    Forest Service reorganization 188
    Pulpwood Agreement No. 16 233, 234–37
    regulations 139, 140, 141, 143, 145, 188
    scaling logs 111–12, 113–14
    timber allocation 62, 90–91, 94–97, 141, 143, 188
British Columbia Railway. *See* BC Rail, PGE
Brown, Don *208,* 250
Bruce, James 86
Buis, Leo 154
Burns Fry 241
bush camps 56–57, 58, 60, 69, 71, *71*
butt 'n top loader 141, *141*
Butterfield, Jim 245
Byers, Dawn-Ann 243

Cambio 26-inch debarker 120
Camp 3 *25,* 26, *28,* 39, *40,* 41, *43*
Canadian Car Company (CanCar) 125, 129, 130
Canadian Forest Products (Canfor) 120, 121
Canadian market downturns 59, 201, 203, 239–40
Canadian Standards Association 283
Canfor. *See* Canadian Forest Products
Canim Lake Sawmills 85, 94, 97, 100
Cargnelutti, Eugenio *192,* 197
Cariboo
    aviation 159, 163
    consolidation of logging industry 97, 125
    forest industry expansion 87
    history 86, *88*
    land-use strategy (CORE) 247, 250
    logging 52, 61–62, 82, 97, 100, 114–15 (*see also* Edwards Lake site, Meadow Lake)
    logging roads *142,* 143, 145, 154–55
    timber supply *228*
Cariboo Lumber Manufacturers Association 112, 116
Cariboo Manufacturing 63, 75–77, 100 (*see also* Kardos, Laci)
Cariboo Woodworkers Association 201–5
Carlson, Ray *244*
Cathro, Doug *133*
Cattermole planer 147
cattle rustler 171
Cawley, Con 133
Cecil, Lord Martin 86, 87, 88
Cessna airplanes *159, 160,* 161–62, *163,* 182
    jet 165, *166, 167,* 180, *180, 182*
chainsaws 27–28, *29, 30,* 31, *32,* 33, 46, 48–50
Chamiss Bay 41–42
Chasm 52
Chasm division 147, *148, 151, 154,* 193, *201,* 214

beehive burner 152
construction 148–49
converts to stud mill *152,* 153–54
gasification plant 150–52
1982 upgrade 205
1995 upgrade *267,* 270
1997 upgrade 269–70, *271*
problems 149–153
sawmill sold 275
startup 149–50
timber supply 147, 153, 188, *189*
treated lumber 147, 152–53, *153,* 189
wood chips 155
Cheston, Wes 236, 247
Chilliwack 42
Chip-N-Saw 125–26
problems 130–32
production level 132, 134
chipper 121
Christen Husky *164,* 179, *179*
Clark, Joe *227*
Classen, John 129, 215
clearcut logging 143
Climax locomotive 37–39, *43*
Clinton 146–47 (*see also* Chasm division)
Clinton Hotel 55
Clinton Sawmills 52–53, 59
Code, John *154*
Cohen, Dave 239
Commission on Resources and Environment (CORE) 247, 250, 266
Cook, Dave 255, 260–262, 272, *284*
Cook OSB division 277, *281*
Cooper, Arnold *86*
Cornthwaite, Wilbur *154,* 206
Cotter, Chester 91–93, *92,* 94, 100
Cotter, Jerry 120
Council of Forest Industries 212
Cowichan Valley, logging 25–27, *31,* 34–37, 39
CP Air Flight 21 172–73
Cretin, Bill 66, *67,* 68
Crispin, Cliff 104
Cultus Lake 42, 53, 70, 73

Danczak, Joe 78–79, 94, *96,* 144–45, 181
Dast, Guenther 78, 83, *117*
David and Susan Ainsworth Fund 283, 284
Davidson, Bill 223
Dawson Creek 228, 255
Day, Bill 62, 147
Daykin, Jerry *194*
delimbing 106, 136, 138, *138*
Depression
Alberta 18, 20, 22
B.C. 25
Dewitt, Don *86*
Dickson, Dick 129, 143, 149, 199, *199*
Dog Creek 164–65, 171
Dohman, Grant 136
domino falling 73, 108
Douglas fir *26, 29, 30*

in Cariboo 55, 61–62, 87, 91
harvesting 104, 108
scaling 110, 111–12
studs 92, 94
Downie, John 167, 183
Drott, Erv 136
Drott 40 feller buncher 136
drying kilns 101–3, 132, 150
Duckworth, Jack 221, 222, 223
Duncan, Larry 154–55, 207
Durastrand 281

Edwards Lake site 69–75, 77–78
Elmendorf, Armin 229
environmentalism 188, 223, 239–40
Ernst, John 125
Ernst Lumber 125
European market 104, 266
Evans Forest Products 217–19
Eversfield, Art *86*
Exco Industries 148, *148,* 210, 221, 223
Exeter Road 63, *72*
Exeter Road site 78–79, 84, *193*
beehive burner 85, *93, 102,* 103, 121, 126
finger-joining mill *190, 191,* 191, 193, 214–15
gang saw *87, 92*
generating centre 85
hydroelectric power 85–86
log yard crew *239*
long-time employees *206*
moving in 81–83, 85
1959 *90*
1960s *98–99, 127*
1970s *130*
1990 *266*
office 84, *127,* 157, *190* (*see also* Ainsworth Lumber)
sawmill closure 269, *269*
stud mill 103, *125,* 126, *128,* 128–32
wood chips *121*

Federal Business Development Bank 241
Fellenz, Hanns (son-in-law of DA and SA) *208,* 210, 285
Fellenz, Ryan 285
Fellenz, Steven 252, *253,* 285
Fellenz, Susan 285, *285*
feller bunchers *135,* 136–37, 139, 223
Fibreco Exports Inc. *122,* 123, 155–56
Findlay, Stan 85
finger-joined lumber 189–93, *192* (*see also* Abbotsford division, Exeter Road site)
fire fighting 170, 181, *181*
First Nations
Grande Prairie OSB plant 256, 257, 262
High Level OSB plant 272, 274
land-use decisions 224, 250, 268
support AL 236
Fish, Dick 161
Footner Forest Products 273–74 (*see also* High Level OSB plant)
Forest Act (1978) 187–88
Forest Care standard 283

forest industry 69
    dangers 34–35, 36, 46–47, 108, 139
    safety equipment 34, 109
    technological changes 25, 27, 36, 119–20, 123, 125
        computerization 209–10, 270
        cost 103, 120, 128, 138, 267–68
        debarking and chipping 120–21
        employees' learning curve 132, 136
        forklifts 66, 68, *68*
        harvesting 28–33, 46, 48–50, 104, *134*, 134–37 (*see also* butt 'n top loader, chainsaws, delimbing, feller bunchers, roadside logging, skidding logs)
        locomotives 37–39, *37*, *43*
        oriented strand board 225
        portable sawmills *47*, 47–48, 62, 104, 120
        road building 143–45
        sawmill design 126
        saws 69, 125–26 (*see also* chainsaws)
        scaling logs 110–14
        trucks 31, 109
        wood chips 122–123
        yarding systems 36, *45*, 104, 267, *268*, *273*
    waste 62, 69, 119–20, 126, 151, 270
    See also specific province or region
Forest Management Agreement 274
Forest Practices Code 266–67
Forest Renewal Act 266 (*see also* Forest Practices Code)
Forest Renewal B.C. (FRBC) 267
Forest Stewardship Council 283
forklift 66, *67*, 68–69
Forman, Mickey 223
Forsyth, Heather *284*
Fort Macleod airport 22
Fouquet, Robert 212, 275
Fowler, Amos 52
Fraser, Don 136, *149*
Fraser, Russ *89*
Fraser Canyon road 54–55, *70, 73*
Friend, Albert 52–53, 56, 59–60, 62, 69
Friesen, Ron 223

Garrow, Cliff 64, 65
Garrow, George 64, 65
gasification plant 150–52, *150*
Georgia Pacific 236
Glover, Bob *133*, 136, *194*
    Exeter Road site 129, 132,
    Exeter site manager 198, 206, 269
Gordie White, 223
Gordon, Jack and Doris 41
Gordon Capital 241
Gosselin, Omer *208*
Grand Rapids OSB division 277–78, *279*
Grande Prairie 254–55, 284
Grande Prairie OSB plant 258, *259*, *260*, 263, 275
    bid 254–58
    construction 258, 260
    expansion 274, 279
    staffing 260–62, *261*
    startup 261, 262–63, *262*
Grant Forest Products 273 (*see also* Footner Forest Products)

Green, Gordon 243
Greenall, Ken 126, 132, 148–49, 157, 207, *211*, *213*

Halcro, Mildred 71
Halcro, Stan 53, 59–60, *59*, 62, 74, *86*
    moves bush camps 71, *71*
    runs DA's portable mill 77
    supplies timber to AL 91, 105, *106*
Halkirk 14–15, 17
Halkirk Literary and Debating Society 19
Hamilton, Ian *256*
Hamilton, Millie 184
Harcourt, Mike 239
Harris, Tyler 245
Harrison Lake logging 42, 45–46
Hess, Merv 167
Higgins, Jim 108, 139
Higgins, Marvin 108, 139
High Level OSB plant 272–74, *273*, *274*
Hihium region 155, 218
Hopkins, Dave 122
Hopkins, Mark 250, *256*
Horsefly 100
Hunt, W.E. 152

Industrial Development Bank 103, 120, 128, 241 (*see also* Federal Business Development Bank)
Industrial Engineering Ltd. 33, 49–50
Industrial Timber Mills Ltd. (ITM) 25, 25–26, 36 (*see also* Camp 3, Cowichan Valley)
Inland Empire 92, 93
Inman, Red 78
Interex Forest Products 281
International Woodworkers of America (IWA) 36, 51–52, 201, 205
ISO standards 282–83

Jackson Lumber Harvester (portable sawmill) 11, 47–48, *47*, *48*, *50*, *51*, *54*
    in Cariboo 52, 55–56, 63–64
Jackson, Clinton D. 47
Jacobson, Harold 100, 112
Jacobson Bros. Sawmills 100
Jacobson, Odt 100
Jamieson, Howie 167
Japanese market
    OSB 234, 238–39, 274, 280
    standards 195
    studs 104, 195–96
Jens, Rudy 97, 100
Jens, Slim 97, 100
Jens, Walter 97, 100
Jones, Lyle 129
Jones Lake Logging 51
Judson, Louis 217

Kardos, Laci 59–60, 62–63, 75, 100
Kardos Forest Products 100 (*see also* Cariboo Manufacturing, Kardos, Laci)
Keller, Jim *194*
Kelowna Machine Works 66

kerf 126
Kerr, Leslie 59, 100
Ketcham, Bill 100
Ketcham, Pete 100
Ketcham, Sam 100, 131, 174
Kite, John 167
Klein, Colleen *284*
Klein, Ralph 258, 272, *284*
Koffman, Morley 219, 221, 241, 243
Komori, "Fuzzy" *105*
Komori Lumber 100
Kopec, Cas *89*
Kowalski, Ken 258
Kozakevich, Met 198
Kraft, Ed *206, 208*
Kraus, Charlie 129

L'Heureux, Larry 136, 139, 140–41, *144*
Lake Logging Ltd. 41
laminated strand lumber 282
Lampman, Marg 223
Lancaster, Gordon 243
LaRue, John 223
Law, Steve 250
Leach, Mike *250*
Leary, Carl 272
Ledcor Industries 243
Lee, Rod 155, 194, 197, 205, 240, *243*
Lefferson, Alan *79*
Lefferson, Jack 55, 73, *76*, 78, 79, *201, 206*
Lignum Ltd. 59, 100
Lillooet division 217–19, *217, 218, 219*, 265
    Ainsworth purchase 222–24
    environmental groups 223
    Forest Practices Code 267–68
    problems 223–24
    sawmill closure 269
    timber supply 219, 222, 223, 250
    veneer plant upgrade 270, *271*, 271–72
Little Bridge Creek Logging 107
lodgepole pine
    in Cariboo 91, *228*
    Clinton timber sale 146–47
    delimbing 106, 138
    drying 91, 93, 101–3
    harvesting methods 104, 107–8
    increased harvesting 135–37, 139
    milling 108
    mountain pine beetle 188, *189*
    OSB 230, 231–32, *231*, 251
    pressure treating 152
    problems with 91
    quota 94–97
    scaling 110–14
    skidding logs *96*, 108–10, 137
    stud production 103
    studs 91–93, 94
    supply 94–96
    wood chips 252–53
Loeppky, Paul *149, 206*
logging trucks *31*

improved performance 200, *200*
loading 55, 64, 65, *65*, 66, *76*
radio communications 201
Lokomo felling head 223
Lone Butte 100
Lone Butte mill 156–57, 205
Louisiana-Pacific 255

Magnuson, Blair 171, 197, 240, *243*, 280
Mahoney, Stan 31
Malkin, Bob 63
Marathon gang saw 69, *75, 76*
Marathon II gang saw 82, *84, 92*, 103
Marks, Ross *89*
Marshall, Stu 129
Martin, Pamela 204
McCloskey, Kelly *235*, 236, 237, 246, 250
McCormack, Al 206
McCormack, Bob *135*, 198, 200–1, *208*
McDuff, Betty 55
McDuff, Jim 54–55, 57
McGladdery, Don 191–92, 215, 216
McGladdery, Sue 192, 216
McGregor, Irwin *49*
McIntosh, Len *133, 149*
McKim, Art 22
McMillan, Clarence 129
McMillan, Glenn 100, *105*, 156
McMillan, Jim 100, 156
McMillan, Vern 129
McMillan Contractors 100, 156, *156*, 205
McNabb, Gordon 139
McNabb, Merrill 139
Meadow Lake 53, 55–59
metric system 146
Mickelson, Ingvald *35*
Miller, Jim 245, 246, 250, *250*, 251
Minhas Bros. Trucking *262*
Monette, Dowling 139
Monteith, Melvin *86*
Moore Canada Ltd. 150
Moore drying kilns 101–3
Mori, Frank 223
Moroz, Ed 223
Morrissette, Albert 223
Morter, Doug 133
Morton, Jeremy 184
mountain pine beetle 188, *189*
Mountain Pine Lumber Ltd. 133, 155, 196
Mt. Timothy Ski Hill *225*

Nadin, Pete 104, 106, 118, 129, *206*
    AL forester 143, 147, 188
    Chasm mill 149
    flying with DA 159, 170, 171
Nadin, Robin 104, 106, *208*
    Bonaparte River bridge 155
    building roads 144
    flying with DA 184, 185
    logging foreman 140
    moving caboose 215, *215*

moving Cats 198, 207
Nelson, Harry *86*
Newman planers 73–74, *74,* 75–77, 81–82
Niemiec, Ed *149*
Norton, Al 171

Olson, Carl *35*
108 Mile airport *166,* 167, *178*
108 Mile Heritage Site 11
100 Mile House
   airstrip *161*
   early phone service 69
   history 86, *88–89*
   incorporation 87, *89*
   1950s *60,* 61, 86–87
   1960s *89*
   1980s recession 202–4 *204*
   support for AL 204, 234, 236, 237, *238*
100 Mile House Lions Club *86*
100 Mile House school 62
100 Mile Lodge *88*
100 Mile OSB plant 234, *248–49*
   construction *242,* 243–44, *244, 245, 246, 247,* 251
   first press load 246, *251*
   production 275
   staffing 244–45
   startup 251–53, *252, 253*
   timber supply 230–33, 246–47, 250–51(*see also* Pulpwood Agreement 16)
103 Mile site 60–61, *61, 62, 63,* 63–69
oriented strand board (OSB) 225, 229, *229,* 232, *274*
   introduced to North America 227–28
   market slump 274–75
   timber supply needed 230–31, 232–33 (*see also* Pulpwood Agreement No. 16)
   See also Ainsworth Lumber, Alberta, *specific plants*
Ostby, Ray 247
Owen, Al 227–28, *227,* 231–32

Pacheco, Joe 203–4, *206*
Pacific Coast Militia Rangers 41
Paterson, Walter 81
Pearse, Peter 187–88
Peebles, Steve 215
peelers 52
Pelican Spruce Mills 228 (*see also* Owen, Al)
Pelkey, Carl 78–79, 94, 100
PGE 78, 83, 197
Phaneuf, Bob 147, *208,* 225
Pigeon Creek Turkeys *208*
pine. *See* lodgepole pine
Pinette, Conrad 118, 133, 156, 175, 242
Pinette, Gabe 94, 112, 118
Pinette and Therrien Mills Ltd. 93–94, 100, 112, 133, 190, 196
Pinkney, Anne 107
Pinkney, Larry 107
Plewes, Lynne 157
plywood vs. OSB 225, 227, 229
Poirier, Leo 152
poplar 272, 283
Porter, Ma 56, 57

Porter, Matt 56, 57
Potlatch Corporation 277
Pourform panels 218, *224*
Price, Val *49,* 55
product certification 191
pulp and paper mills, in B.C. Interior 120–21
Pulpwood Agreement No. 16 233–237, 246–47, 250
Pulpwood Area Harvesting Agreement 230–31, 232

Quesnel 100
Quirin, Roger *86*

railcars, loading 83, *119,* 121–23, 152, 197–99, *197*
Ramsay, Scotty 165, 169, 176
RBC Dominion Securities 241
RCMP helicopter 164
Reid, Carl 47, 52
Richards, Steve *63*
Richmond, Claude 203
road building *142,* 143–45, 155, 266
roadside logging 139–41, *140, 141*
Robert E. Malkin Lumber Company 74
Robinson, Jim 111, 114
Rogers, David 241
Rohman, D. *206*
Roose, Kimberly 183
Roose, Leo 183
Roose, Vicky 183
Roose, Willie 183
Rowland, Sadie Dobson (mother of SA) 21
Rowland, Thomas (father of SA) *20,* 21–22
Royal Commission on Forest Resources 187
Ruckle, Cec 52
Runyon, Ernie 125
Ryan, Tim 246–47, 254, 255–56, *256,* 258, 272

Sato, Ken *127*
Savona division *216,* 217–19, *222, 224,* 265, 276
   plywood mill upgrade *220,* 221, 270–71, *271*
   sustainable forestry 283
   timber supply 218
scaling logs 110–14, *113*
Schaff, Tom *166,* 167, 180
Seaboard Lumber Sales Ltd. *194,* 195
Second World War 22, 26, 28
Sellars, Dick 112, *194,* 206, *206*
   Chasm site manager 151
   finger-joined lumber 190–92
   flying with DA 181–82
   Japanese market 195–96
   joins AL 115–17
   quality 197
   Susan and AL 118
Sellars, Richard 115
Semley, Ron 78, 83
70 Mile House 56
Shanks, Boyd 223
Sharp, Mitchell 175–76, *177*
Shaw-MacLaren, Chuck 73, 85, *86,* 89, 104
Shaw-MacLaren, Dave *149*
Shewring, Peter 165, *166,* 167

Side, Doug 219
Silveira, Steve *208*
    first "outsider" hired 207–9
    flying with DA 182
    Grande Prairie OSB 254
    Lillooet division 224
    OSB 230, 231, 232, 233, 234
    PA-16 hearings *235*, 236–37
skidding logs 108–10, 137
skyhook *45*
Slocan Forest Products 197
Smedley, Brian *167*
Smith, Al *127*, 132, 206, *206*
    AL employees 117, 118
    Chasm mill 148–49, 153, 205
    finger-joining mill 191
    hard hats 109
    joins AL 115
    operations manager 193
    Savona plant 221
    stud mill upgrade 129
    wood chips 120–21, 123
Snowden, Danny 129
softwood lumber 205, 223, 267, 269
spar poles *33*
Sposato, Luigi *208*
spruce
    in Cariboo 87, 91, 104
    scaling 111, 112
    studs 92, 94
St. Pierre, Paul 175–76, 177
Stetson-Ross Machine Co. 150
Stewart, Ken 203, 206
studs 91–94 (*see also* Japanese market, lodgepole pine, United States market)
Sunrise Lumber 182
Sustainable Forest Management 283
Sustainable Forestry Initiative 283
Svedt, John *29*
Szauer, John 230

Teichgrab, Bob *200*
Teichrob, Len 216
telephone service 69
Teslo, Maurice 62
Theno, Doug 160
Thermastrand 281, *282*
Therrien, Dollard 94
Therrien, Roger 94
timber auctions 90, 95, 278
Timber Sale Harvesting Licence 141, 143
Timber-Toter. *See* forklift
Tolko 274
Tomlinson, John *206*
transit sales 198
transport trucks 198, 199, *199*
transportation. *See* BC Rail, logging trucks, PGE, railcars, transport trucks
Travers, John *49*
Trusler, Gordon *86*, 116

United Kingdom market 51, 104, 195
United States forest industry 278
United States market 195
    boycotts 266
    finger-joined studs 193
    lodgepole pine studs 92–93, 100, 154
    softwood lumber 205, 223, 239, 267, 269
UBC wood science program 212
unemployment insurance stamps 84

Valleyview 279
Vander Zalm, Bill 223
Viette, Ken 223
Vilac, Barry *206*

W.F. Gibson and Sons Ltd. 41
waferboard 229
Waines, Ken 174
Watson, Bob 115, 129
Watson, Carl 69, 82, 83, 102, 149, 197
Weier's Sawmills 100, 133, 196
Weier, Herb 100, 133, 197
Welbourn, Derek *253*
Weldwood 52, 100, 208
    Clinton timber sale 147
    lays off sawmill, planer workers 203
    plywood plant 221, *220*
West Fraser Timber 100, 131, 174, 275
Western Forest Products Lab 191
Western Plywood 100 (*see also* Weldwood)
Westwood Polygas Ltd. 150
Weyerhaeuser 232, 255, 257–58
Whalley, Ernest *86*
White, Doug *146*, 208
    Bonaparte River bridge 155
    flying with DA 159, 184, *184*
    OSB timber supply 250–251
    woodlands 147, 233
White, Gordie 223
Williams, Bob 147
Williams Lake 100
Williston, Ray 188
    and David Ainsworth 96–97
    OSB 227–28
    phases out timber auctions 94–95
    reduces harvesting quotas 90–91, 96
    sets new logging standards 120
wood chips 120–123, 155 (*see also* Fibreco Exports Inc.)
Wood, Norm *154*, 206
Workers' Compensation Board 109
Wright, Dave 243, 244, 245, 258, 260, 263

yarding logs *45*
Youbou 25
Young, Hal 215, 221

Zirnhelt, David 244, *244*